NEW DIMENSIONS IN BIOETHICS
Science, Ethics and the Formulation of Public Policy

NEW DIMENSIONS IN BIOETHICS
Science, Ethics and the Formulation of Public Policy

edited by

Arthur W. Galston
Molecular, Cellular and Developmental Biology
Yale University

Emily G. Shurr
Institute for Social and Policy Studies
Yale University

KLUWER ACADEMIC PUBLISHERS
Boston / Dordrecht / London

Distributors for North, Central and South America:
Kluwer Academic Publishers
101 Philip Drive
Assinippi Park
Norwell, Massachusetts 02061 USA
Telephone (781) 871-6600
Fax (781) 681-9045
E-Mail <kluwer@wkap.com>

Distributors for all other countries:
Kluwer Academic Publishers Group
Distribution Centre
Post Office Box 322
3300 AH Dordrecht, THE NETHERLANDS
Telephone 31 78 6392 392
Fax 31 78 6546 474
E-Mail <services@wkap.nl>

 Electronic Services <http://www.wkap.nl>

Library of Congress Cataloging-in-Publication Data

New dimensions in bioethics / edited by Arthur W. Galston, Emily G. Shurr.
   p. cm.
   ISBN 0-7923-7249-2
   1. Bioethics. I. Galston, Arthur William, 1920- II. Shurr, Emily G.

QH332 .N49 2000
174'.957--dc21

00-049322

Printed on acid-free paper. Printed in the United States of America

The Publisher offers discounts on this book for course use and bulk purchases. For further information, send email to <molly.taylor@wkap.com>.

Contents

PREFACE

The birth of this collection of essays was serendipitous. In 1993, Bradford H. Gray, then Director of Yale's Institution for Social and Policy Studies (ISPS), invited me to initiate a graduate seminar in bioethics and public policy at ISPS, modeled after my long-running undergraduate seminar on a similar subject. Each academic year since then, the seminar has invited approximately nine speakers, one per month, to address a group of interested faculty, postdoctoral fellows and graduate students on a topic of bioethical interest. These short formal presentations invariably spawned a spirited discussion period for at least an hour afterward. When Donald Green became the new Director of ISPS in 1996, he approved continuation of this seminar while also urging that we consider publication of some of the talks. This volume is the result of the willingness of the selected authors to prepare manuscripts based on their presentations at the ISPS seminar.

I am especially grateful to Brad Gray and Don Green, who first suggested and supported the collection of essays for a publishable volume, and I am indebted to Carol Pollard, Coordinator at ISPS, for insightful and efficient execution of innumerable tasks associated with preparation of the material for publication. Emily Shurr and I collaborated on the presentation of these pieces as a unified and socially compelling collection.

The Institution for Social and Policy Studies at Yale occupies a unique interdisciplinary niche on campus. A recent seminar included representatives from the Schools of Medicine, Law, Divinity, Forestry and Environmental Studies, and from the Departments of Political Science, Economics, Philosophy, Molecular, Cellular & Developmental Biology, and Ecology & Evolutionary Biology. This tremendously yeasty mix guarantees lively and broad-spectrum discussions and the formation of cross-cultural intellectual contacts and collaborations. More than 20 years ago, with the help of Director Ed Lindblom, I first tested at ISPS my ideas for a bioethics course at Yale. I greatly treasure my continuing association with this Institution.

Arthur W. Galston
September 1999

INTRODUCTION

Coming into its own as an active scholarly discipline in the early 1970s, the new field of bioethics applied its brand of ethical analysis to a growing multitude of vexing problems arising from rapid advances in the biological sciences. For when successful scientific effort emerges from the laboratory into the world of technology, commerce and citizens' everyday lives, it quite often raises questions never before contemplated by society. Many of these questions require analysis by physicians, lawyers, theologians and, ultimately, the public and the lawmakers who write and pass legislation which establishes codes of behavior to guide the use of new science for the benefit of society.

We have constructed this volume in the hopes that it will be useful for several broad audiences. Unlike many "traditional" bioethics texts, however, ours is not a survey of topics in medical ethics. We make much of the scholarly tradition and field of bioethics as it has developed, and we use that sound scholarship as our predecessor and the foundation of our field. But we are hoping to do a new thing with this book: after its initial—greatly successful—thirty years or so, we're pressing the boundaries of the field outward, and in so doing calling for increased cross-disciplinary collaborations. A standard "medical ethics" text would include the usual important categories (physician-patient relationships, equitable distribution of health care services, reproductive issues, informed consent). Those issues are addressed here, but in a new way: as they come into contact with policy and lawmaking. New industries and new medical services call for new government attention to determine adequate regulations with society's best long-term interests in mind.

As citizens, professionals, and students we still need to study those ever-pertinent questions. Indeed, the indispensable volumes by Beauchamp and Childress, Arras and Steinbock, and others are vital now more than ever and it is a boon to society that new editions are produced. We would only caution against allowing the earlier scope of the field to constrain new directions of inquiry. The field is expanding rapidly; it now becomes clear that the traditional methods and lines of inquiry ought to be applied to ever more pressing biological issues of consequence far beyond the laboratory or the operating room, and into the physical habitat of all of humanity. But the way we address these questions also has consequences for human culture. Whether we as a global civilization prioritize cash flow over human health, or clean water over this year's corporate profits, is determined in an ongoing series of everyday decisions.

This human culture, civilization itself—the lives we must lead together as a species, in our common (global) habitat—is what policy means to influence. The task of lawmaking is governance—of a national economy, of international relations, of what people do to and for each other. Defining in concrete, specific terms what it means for an individual to have a "right" among other individuals, what it means for a nation to have "sovereignty" among other nations—policy is a process by which we delineate and shape the manner in which we as individuals and as commonwealths relate to each other, define our priorities and create our future.

It is essential in any interdisciplinaryeffort for each specialist to learn other specialists' methods and research values in order to reach a mutual respect among disciplines. Certainly the political scientist and the research geneticist can stretch their finely-honed inquisitive minds to relate to one another on an equitable (if not a charitable) basis. We hope that this book will serve exactly that purpose—to give lawmakers a plain-language insight on what philosophers are up to, to give physicians a clear understanding of what international law is about—but moreover, to reveal the surprising degree to which their work is interlinked. Thus, the discussions contained in this book should ultimately serve as an interdisciplinary tool, a bridge-builder between academe and policy, between scientists and humanists, and between the specialties and the general public.

Combining the areas of genetics, medicine, and the environment with legal, historical and policy analysis is central to our concept of "new dimensions." We hope that this synthesis will contribute to changes in the way bioethics is envisioned, integrated and taught, especially at the undergraduate level. We propose this book, therefore, as a useful classroom tool—not to substitute for the indispensable survey texts mentioned elsewhere, but instead as a supplemental reader on new directions in the field and its overlap with policy issues. In the classroom and perhaps the boardroom, the work herein can also serve as the basis for challenging, instructive discussion.

Categories of interdisciplinary inquiry

The three distinct groupings of these essays overlap one another and may provide a framework for discussion: writings especially pertinent to medicine, genetics, and the environment.

Medicine

Before biologists understood how to fertilize human eggs *in vitro*, society did not have to grapple with ethical and legal problems associated with the production of "test-tube babies." But when Steptoe and Edwards described how to do so in 1972, and the first test-tube baby, Louise Brown, was born in 1978 in Lancashire, England, she immediately presented a host of perplexing new ethical problems for which humankind was completely unprepared, and with which we are still grappling today. For example, there is still no satisfactory answer to the question of what to do with the "extra" embryos produced by the *in vitro* fertilization procedure. Obviously, not all such supernumerary embryos can be implanted into the womb of a receptor woman, yet considered as potential human beings, they cannot be casually flushed down the drain. To freeze them in the hope of subsequent implantation, as we tend to do now, is literally to "put the problem on ice," and to defer a difficult decision.

Genetics

Before Watson and Crick elucidated the structure of DNA in 1953, and before discovery of techniques permitting the splicing, manipulation and transfer of DNA from one organism to another, the ethical consequences of mapping a human genome or of genetic engineering could not concern us. But today, we must ponder such questions as the desirability of genetic alterations of human cells for therapeutic purposes, and the evolutionary and ecological consequences of genetically engineered crops.

Environment

Before Rachel Carson's <u>Silent Spring</u> made us aware, in 1962, of the adverse ecological consequences of the widespread use of pesticides like DDT, we were not as ethically concerned about the disappearance of a species, the maintenance of biological diversity and the preservation of sound ecological conditions as we are now. In some countries, new political movements centered on "Green Parties" have arisen to influence societal policy with regard to the environment.

Thus, as a result of its application to a multitude of problems in the last three decades, bioethics has found wide use in academic, medical, legal, business, governmental and administrative circles, and in the process, has itself undergone considerable evolutionary modification.

As Albert Jonsen recounts in <u>The Birth of Bioethics</u> (Oxford, 1998), *medical ethics* began before bioethics. The agonizing problems that challenged the family doctor and the surgeon required decision making that transcended the ethical skills of many practitioners, and compelled collaboration with theologians and philosophers more specifically trained in ethics. Ethicists, on their part, sensed the need for an integrated and cooperative approach to their analysis of complex scientific problems. Because the growing interdisciplinary nature of the subject forced such collaboration among research scholars of all kinds, the ranks of those interested in solving bioethical problems came to include many natural scientists without formal ethics training. Such newly-hatched bioethicists have sometimes been criticized by philosophers for "practicing ethics without a license." The newcomers have responded that just as they have had to struggle to learn how to apply ethical analysis to issues arising out of biology or medicine, so philosophers must labor to understand the intricacies of biochemistry, genetics and medicine as a prelude to informed ethical interpretation in those fields. Indeed, in many bioethical analyses, a primary task is to get the scientific facts straight. In these days of rapid advances in biology, this is sometimes no small problem. Accordingly, several essays in this book deal with scientific complexity by using extended documented reviews to guide ethicists and other "lay persons" unfamiliar with the technology or biological process under discussion.

The generous support of biomedical research, together with the extraordinary poignancy of many urgent medical problems—whether to save a tragically deformed newborn, whether to withdraw life support from a comatose patient, whether to use patients in potentially risky medical experiments—create an urgent demand for some sort of ethical guidelines. Aware that serious ethical lapses have occurred (as in the infamous Tuskegee syphilis "experiment") and could easily occur again, physicians and ethicists work together to arrive at procedures that forestall and hopefully prevent such occurrences in the future. Their efforts have led to the formulation of a series of rules to guide the physician and protect the patient.

At present, the protocol for any proposed experiment involving human subjects must be approved in advance, either by an Institutional Review Board at a hospital or a screening panel at a federal agency such as the National Institutes of Health. For advice on such matters as the use of human embryos in experimentation or in cloning, the President of the United States found it expedient to depend on advice from a National Bioethics Advisory Commission. The preeminent role of medical ethics came to dominate the emerging field of bioethics and, in the understanding of many practitioners and students, became synonymous with the entire field.

Examples from genetics.

Bioethics is no longer exclusively concerned with medical problems. Avery, McLeod and McCarty had discovered in the 1940s that the genetic material in the *Pneumococcus* bacterium was not protein, as popularly supposed, but most probably DNA. In 1953, when James Watson and Francis Crick proposed a structure for the DNA molecule, it became clear almost immediately that their structure explained not only how genetic information for all creatures is stored in and retrieved from DNA, but also how the binary nature of the molecule guarantees exact transmission of replicates of this information into subsequent generations of cells. With this understanding, it was not long before a few technological advances made genetic engineering possible. Now, specific information-carrying bits of DNA, the genes, can be located in the giant DNA chain, precisely excised from the chain, incorporated into "vectors" that carry them into bacterial cells, and there replicated many times. These cloned genes can then be harvested and in turn introduced into other types of receptor cells, including those of humans, where they can often be activated to control new cellular characteristics. From a medical point of view, this seems to offer tremendous possibilities, for if a patient has a biochemical deficiency caused by an ineffective "mutant" gene, then one might correct the abnormality simply by introducing the normal version of the gene.

Obvious ethical questions arise here. For example: while most people can approve the correction of a chemical abnormality in a sick patient, many instinctively recoil against transforming the reproductive cells, the sperm and the egg. Since the gene transferred would cease functioning when that individual died, this type of gene therapy does not seem to differ much from the administration of a medication like an antibiotic. But introducing that same gene into reproductive ("germ-line") cells, from which it could be passed on to offspring, raises the specter of eugenics as well as of starting a line of genetic transmission that might be difficult or even impossible to stop. This would be wrong, according to some critics, because it seems to involve "playing God," initiating a process that might adversely alter the course of human evolution.

Is this type of problem simply a question for a "medical ethicist" to analyze? If so, is genetic ethics simply a subset of medical ethics? The answer, surprisingly for some, is in the negative. It turns out that recombinant DNA techniques are valid not only for humans, but for all organisms, including microbes and agriculturally important plants as well. In fact, naturally-occurring genetic engineering was first discovered during a study of a plant disease called crown gall. The invading organism, *Agrobacterium tumefaciens*, caused a malignant tumorous malformation to appear because it transferred a piece of its DNA containing a "tumor-inducing principle" to a cell of the plant. This bit of *Agrobacterium* DNA (appropriately "disarmed" so as not to cause tumors) can carry almost any other gene with it when it enters a plant cell. It is now used routinely as a vector for transferring desirable genes into receptor cells. Of all the transgenic organisms engineered to date, plants predominate by far.

The widespread use of plant engineering has caused concern about possible ecological and evolutionary consequences of transferring genes into organisms that have never had them before. What might the consequences be, for example, of introducing a gene for cold-hardiness into a previously cold-sensitive crop plant? Might pollen grains carrying this new gene spread it to wild relatives of the crop plant, creating hardy hybrid weeds capable of invading new areas, thereby producing agricultural and ecological havoc? Clearly, genetic ethics extends far beyond the medical arena and forms a separate subsection of bioethics, with its own specific set of ethical dilemmas.

Examples from environmental ethics.

For over three decades, stimulated greatly by Carson's *Silent Spring* and the work of other scientists and activists, we have become aware of the many deleterious effects that some aspects of human technology produce on the natural world around us. This new awareness has stimulated considerable study of the presumed conflict between economics and ecology; by now, the field of environmental ethics has established itself as a legitimate section of bioethical study. Many profound ethical questions have been raised and pondered, including: What is our ethical responsibility for the maintenance of species diversity? Of stable ecosystems? Of renewable resources? Of sustainable agriculture? How many people can the earth support if its present productivity is maintained? What kind of living standard is implied in the answer to that question—that of the present United States, of Bangladesh, of something in between? Is it possible to abolish hunger through increased agricultural productivity? Why should we be concerned by the massive and growing use of persistent pesticides? What ethical dilemmas surround the industrial emission of sulfur dioxide and the subsequent fall of acid rain that depresses crop yields in distant locations? What do we mean by a "land ethic" and what moral limitations does it place on our behavior?

The process of addressing these questions helps determine the fate of *Homo sapiens* on our planet. Clearly, then, environmental ethics also qualifies as an important division of bioethics. In presenting this volume we recognize the importance of all three subjects (medicine, genetics, environment) when we speak of "new dimensions in bioethics." By bringing these essays together in a single source, we hope to influence the way the subject is researched and taught.

Forging agreement in bioethics.

How has bioethics arrived at its accepted canons of rules and procedures? Agreement among scientists almost always depends on independent replication of a published experiment or, in some cases, on confirmation of a theoretical analysis. Agreement among ethicists, by contrast, depends on mutual acceptance of shared values. This difference is sometimes encapsulated in the statement that "Science tries to tell us how the world is constructed, while ethics tries to tell us how the world ought to be constructed." It can be difficult to harmonize these two modes of thought and operation, and it should therefore come as no surprise that bioethical analysis seldom offers hard-and-fast solutions to practical problems. What it contributes instead is a methodology that aids in systematic considerations and analysis of ethical problems created by technological applications of biological research (sometimes by reference to classical ethical theories.

Organization of this volume

The discussion begins with a provocative essay by Nobelist **Joshua Lederberg**, who considers the social function of the scientist in the light of J.D. Bernal's analysis of that subject in 1934. Lederberg asserts that the scientist's preeminent responsibility is the preservation of the integrity of science itself by doing good work. He notes the frustrating role of the scientist as consultant and adviser, especially when these activities bring him or her into politics. The fates of Andrei Sakharov and J. Robert Oppenheimer, authentic scientific heroes, bear witness to the sometimes unfortunate interplay of politics and science.

In initiating the section on genetic ethics, **Kenneth Kidd** reports his findings on the comparison of human chromosomes from various racial, geographica and ethnic groups. Using molecular techniques for comparative analysis of DNA structure for particular groups, he concludes that the human population has a unitary African origin, and that because of massive gene flow among major groups, it is scientifically inaccurate to divide us into distinct racial categories. It is also inaccurate to speak of a single "human genome," since we all differ from each other because of constantly occurring mutations. The human genome project will thus describe a pattern around which there is considerable variation. The consequences for social policy are profound.

Ellen Messer and **Nina Dudnik** contribute an essay analyzing the ability of the new agrobiotechnology to help solve the problem of world hunger. They find that for social, economic and political reasons, it is not likely to produce a major effect on the current world situation. The essay prompts a great number of questions regarding the ethical use of technology, the ethics of the relation between commerce and the public health, and other related areas.

In the final essay in this section, **Dorothy Nelkin** cautions us against a genetic determinism arising from the newer knowledge of DNA. She decries simplistic statements such as "Our fate is in our genes," reminding us that it is a long path that leads from genotype to phenotype, and that social context and political ideology influence scientific theories. She emphasizes the danger of authentic science being co-opted by politics, commerce, and mass culture. Claims that DNA-based predictions will have therapeutic value have been justified in only a few cases, and extending predictions to behavior may be based less on science than on popular mythology. The social consequences of such predictions may, however, be profound. To a remarkable degree, genetics research seems to influence the social definition of a person in our culture.

Since much of recent medical practice, especially in the field of reproductive medicine, is circumscribed by legislation and judicial opinion, it is appropriate that legal scholar **Robert Burt** takes us through an analysis of the origins of the *Roe v. Wade* decision. Surprisingly, he characterizes this historic event as counter-revolutionary, chronicling how and why each of several influential Justices came finally to vote for Harry Blackmun's compromise formulation. While defining acceptable practice during each of the trimesters of pregnancy, *Roe v. Wade* has not resolved the basic philosophical dispute between "pro-life" and "pro-choice" camps. Too, Burt presents the ethical issues with careful insight—for example, the moral status of fetuses and the seemingly "insoluble conflict" between the interests of an individual woman, a fetus, and society.

Carol Levine takes us from the world of legal analysis to the gritty realities of home care for family members unable to cope with daily maintenance routines by themselves, as well as those families of patients requiring round-the-clock medical attention at home. Examination of various ethical concepts behind the responsibilities of family, private agencies and public agencies permits her to formulate recommendations for an ethically-based system of "nursing" coverage. With special emphasis on the "so-called private choices which have been triggered by public policy," she presents a nuanced, compassionate, practical study of the expensive, persistent problem of end-of-life care.

William Galston, a former member of the domestic policy team in the Clinton White House, then takes us through the behind-the-sciences activities that led to the formulation of reasonable and ethically-based regulations regarding the use of human embryos in research, and how politics influenced the final decision. This question, which gained urgency only after *Roe v. Wade* and the development of *in vitro* fertilization techniques, is still with us, for the necessity of budgetary compromise compelled the President to reject the "middle ground" solution proposed by his advisers. Portraying the process of determining the allocation of public funds on controversial projects, as well as (once again) defining the moral status of human life, Galston's up-close and personal account displays the complex interaction among social philosophers, theologians, politicians, the White House, a major government research agency, and scientists to form public policy for our nation.

Jay Katz concludes this section with a chilling examination of the role of Nazi doctors in the experiments conducted on human prisoners in the extermination camps. He is able to relate their gruesome practices to attitudes widespread among physicians long before the Holocaust, and to place them as part of a continuum of current medical practices. Analyzing the differences between doctors as healers and doctors as researchers, he reveals that medicine has not always sought to obtain prior voluntary consent from patients, and has sometimes used such consent as a subterfuge for shifting moral responsibility from themselves to their patients.

Robert Socolow introduces us into the section on the environment with an analysis of the "moral terrain of ecological vulnerability." He reasons that human impacts on the environment have recently become prominent because of their magnitude, their measurability and our willingness to reason morally about their consequences. The interactions of these three factors of scale, awareness and conscience can work to mitigate continued degradation of the environment.

David Pimentel and **Kelsey Hart** then provide a striking case history in a study of the many positive and negative environmental impacts of the widespread use of pesticides. Against the undoubtedly beneficial effects of pesticides on crop yields, we must quantify their social costs in causing environmental degradation and adverse public health effects. **Arthur Galston** continues with a description of the biology and politics of the use of Agent Orange as a defoliant in the Vietnam War. This specific case history, a very tangible and practically important instance in which bioethics and policy come into contact, once again requires ethical decision-making against a backdrop of conflicting "desirable" military effects and harmful, if unintended, side effects on the environment and public health

Strachan Donnelley analyzes our relation to animals in three contrasting contexts: as objects of experimentation in the laboratory, as household pets and as wild creatures. The visions of Aldo Leopold, Alfred North Whitehead and Hans Jonas are employed to envision the appropriate human-animal relationship in a global macro-view, exploring issues of animal research ethics as well as social-environmental philosophy and its policy implications.

Finally, how do we view nature and what values do we assign politically to features of the environment? **Arthur Westing** concludes this section with an analysis of environmental values embraced by both the civil and military sectors of society. Values are deduced from examination of the Resolutions passed by the 192 sovereign member states at meetings of the United Nations General Assembly and from the various multilateral treaties dealing with humanitarian and environmental principles. Surprisingly, values embraced by our military resemble those held in civil society. Environmental values become more clearly expressed as levels of democracy, probity, wealth and human development improve.

This collection represents a gathering of thinkers, in many cases crossing new frontiers in their own disciplines while at the same time forging new and badly needed links with other disciplines. Presenting them together in this book points the way to an open network of interdependence among traditionally separate modes of inquiry. Furthermore, every author represented here is concerned with bringing bioethical concerns to the attention of policy makers. And all the authors contribute tangibly to an ongoing scholarly effort to inform the public and policymakers about pressing ethical issues involving biological sciences.

Collectively, these essays provide an overview of selected areas of lively current biological ethical and political thought and action with implications for our human aspirations for a better world. Together they call for professional, national and global attention to values and ethics in the public life. The intersection of bioethics and public policy continues to be more and more meaningful to our society. Every issue of the *New York Times* or *Wall Street Journal* brings news of some new development or challenge in medical research, health care distribution, environmental hazards or genetic engineering. We find numerous websites on government's role in regulating the industries and groups most affected by these new fields. Many current questions are about the state's place in serving as moderator, mediator, overseer, or custodian of the public health. Bioethics is no longer confined to the Institutional Review Board or the Hospital Ethics Committee. It's in our Congress, our supermarket, and our stock market. Whether the issue at hand concerns farm workers and pesticide application, crop production and ecological stability, or patients and physician-researchers, the larger issue is always that of the proper relation between science and society.

Arthur W. Galston
Molecular, Cellular &
Developmental Biology
Yale University

Emily G. Shurr
Institution for Social and Policy Studies
Yale University

New Haven, CT
October 1, 1999

The Authors:

Robert Burt is the Alexander M. Bickel Professor of Law at Yale University, New Haven, CT.

Strachan Donnelley has been President and Chief Executive Officer at the Hastings Center, Garrison, NY. He is now head of a new "Humans and Nature" program at the Center.

Nina Dudnik is Special Consultant in Biotechnology at the International Plant Genetic Resources Institute in Rome.

Kelsey A. Hart is in the Department of Entomology, Cornell University, Ithaca, NY.

Arthur Galston is Eaton Professor Emeritus in the Department of Molecular, Cellular and Developmental Biology at Yale University, New Haven, CT.

William Galston is Professor in the School of Public Affairs and the Director of the Institute for Philosophy and Public Policy at the University of Maryland, College Park, MD.

Jay Katz is the Elizabeth K. Dollard Professor Emeritus of Law, Medicine and Psychiatry, and the Harvey L. Karp Professorial Lecturer in Law and Psychoanalysis, Yale University, New Haven, CT.

Kenneth Kidd is professor of Genetics and Psychiatry, Yale University School of Medicine, New Haven, CT.

Joshua Lederberg is President Emeritus and Sackler Foundation Scholar at the Rockefeller University, New York City. He received the Nobel Prize in Physiology or Medicine in 1951 for discovering genetic exchange in bacteria.

Carol Levine is the Project Director for Families and Health Care of the United Hospital Fund, New York City.

Ellen Messer, past Director of the Alan Shawn Feinstein World Hunger Program at Brown University, is now a Visiting Associate Professor at the Tufts University School of Nutrition, Science and Policy.

Dorothy Nelkin is University Professor at New York University, teaching in the department of Sociology and the School of Law.

David Pimentel is Professor of Entomology in the College of Agriculture and Life Sciences at Cornell University, Ithaca, NY.

Emily Shurr is a graduate of the Master of Arts in Religion program (Ethics Concentration) at the Divinity School, Yale University, New Haven, CT.

Robert Socolow is Professor of Mechanical and Aerospace Engineering and Director of the Center for Energy and Environmental Studies, School of Engineering and Applied Science, Princeton University, Princeton, NJ.

Arthur Westing, a forest ecologist who has worked extensively with the United Nations environmental program, is currently a private consultant on matters related to peace, security and the environment.

THE SOCIAL FUNCTION OF THE SCIENTIST: THE ETHICS OF TRUTH-TELLING
Joshua Lederberg

It is appropriate, in considering the social function of the scientist, to begin by quoting at some length from the preamble and highlights of a 1993 Carnegie Commission Report on Science, Technology and Government:

> Government is the complex of institutions, laws, customs and personalities though which a political unit exercises power and serves its constituencies.
>
> Science is the search for novel and significant truths about the natural world. These truths are usually validated by the prediction of natural phenomena and the outcome of critical experiments.
>
> Technology is the instrumental use of scientific knowledge to provide, for example, goods and services necessary for human sustenance and comfort and to support other, sometimes contradictory aims of the political authority.
>
> Scientific expertise and technology have always been valued by government. Weapons and medicines, maps and microprocessors: the products of science are indispensable to successful government. So, increasingly, is scientific thinking. Where but to science can society turn for objective analysis of technical affairs?
>
> The Scientific mind can bring much to the political process. But science and politics are a hard match. Truth is the imperative of science; it is not always the first goal of political affairs. Science can be, [and] often should be, a nuisance to the established order, much as technology often bolsters it.
>
> Moreover, many scientists, lacking the policy skills needed to relate their expertise to social action, are uncomfortable dealing with the political machinery.
>
> A vital responsibility of the expert advisor is to clarify technical issues so that the essential policy questions become accessible to the judgment of the community at large. Yet expertise also has distortions, arising from conflicts of interest, differing levels of competence, peculiarly posed questions, and cultural biases. The discipline of the peer group is the main source of the authenticity of the scientific community.

Science, in fact, cannot exist without a community of scientists, a forum for organized, relentless skepticism of novel claims. Science kept in confidence and inaccessible to colleagues' criticism is no longer authentic. The public rendering of advice and defense of conclusions is of the utmost importance. Nevertheless, advice within the political system must often be confidential. Herein lies another structural contradiction and challenge to the design of organization and decision making.

We must thus establish institutions and processes that enable scientists both to be credible within politics and to remain worthy of the continuing confidence of the larger society. To achieve this dual goal, the first social responsibility of the scientist remains the integrity of science itself.[1]

J.D. Bernal, a scientist already renowned for his experimental work, became a pioneer in the systematic examination of the reciprocal relationships of science and society when he wrote the influential piece The Social Function of Science in 1934. In this article, I will deal with a number of issues bearing on those themes. Despite my doctrinal differences with Bernal's Marxist perspective, and my stress on a U.S. experience versus his in Britain, I have found much to be inspiring and provocative in his writings, throughout most of the 60 years that have elapsed since then.

Had I shared his disillusionment about the automatic benefits to humanity from scientific discovery, I might have been deterred from my initial commitment to a scientific career—one that goes back almost six decades. We recall that 1939 was the dramatic beginning of World War II; it also occasioned the World's Fair in New York City, a celebration of the applied benefits of science—nowhere better epitomized than by the slogan of the DuPont Corporation: "Better Things for Better Living Through Chemistry."

Six years before, Albert Einstein had already had to reverse his prior uncompromising pacifism, and urge the Western democracies to mobilize for the defeat of Hitler. In the U.S., the mobilization of science and the development of nuclear weapons followed from the same imperative of preempting Hitler. The consequent culmination of that scientific and technological effort in 1945 was the division of the world into sovereign, nuclear-armed superpowers. As Einstein tirelessly taught, this undercut any simple ethical or political constructs of the humanitarian consequences of humanitarian advance. It left all scientists

---

[1] Carnegie Commission on Science, Technology and Government. "Science, Technology and Government for a Changing World" 1993.

deeply concerned about their obligations to society. For a generation, it also necessitated that nuclear physics be consulted in the highest levels of discussion of national politics. This has been a precedent for a new relationship of all scientific expertise to government, to the enlightenment of the electorate, and to private conscience.

Bernal commented that Descartes, faced with Bruno's immolation, established the ground rules of the relationship of science to the ecclesial establishment: that these should be mutually incommensurate and non-interfering spheres. That philosophy has endured up to the modern era in the relationship of science to statecraft as well. Bernal writes:

> In his attack on the old philosophy Descartes was as canny as he was courageous. He had no desire to enter into a head-on conflict with organized religion, a conflict that had led to the condemnation and burning of Bruno in Catholic Rome and that of Severus in Calvinist Geneva. He was prepared to be accommodating, and he hit on an ingenious method of doing so which was to make science possible for several centuries at a cost which we are only now beginning to feel.[2]

The separation of religion and science

The effect of Descartes' division ever since was to enable scientists to carry on their work free from religious interference so long as they did not trespass into the religious sphere. This was, of course, very difficult to avoid or refrain from, but nevertheless, it had the effect of producing the type of pure scientist who kept out of fields where he was likely to be involved in controversies of a religious or political kind. To a certain extent Descartes himself must have done this, because the story goes that when he had ready his System of the World, he heard the news of the trial of Galileo and realized that it simply would not do as it was. The Church was clearly determined that the Aristotelian-Thomist system was necessary to secure the truths of the Faith and was not going to tolerate any other system that might put them in question. Descartes consequently set himself the task of showing that his systems could prove the existence of God quite as well as, if not better than, the other philosophies. From his famous first deduction, "Je pense donc je suis"— "I think, therefore I am"—he drew the conclusion that as all men can conceive something more perfect than themselves, a perfect being must exist. Descartes' system was so carefully guarded against theological attack that, in spite of protests from the universities, it was accepted in

---

[2] Bernal, J.D. The Social Function of Science, 1934. See also Bernal, J.D. Science in History, 3rd Edition. MIT Press, 1971.

that most Catholic country, France, within his own lifetime and for a century after his death.

Descartes' system was, however, in spite of its wealth of mathematical and observational content, essentially a magnificent poem or myth of what the new science might be. That was at the same time its attraction and its danger. It was a mixture of conclusions soundly based on experiment with those deduced from first principles chosen, according to Descartes' celebrated Method, only on account of their clarity. The pursuit of that clarity has been the ornament and the limitation of French science ever since. Where in the state of knowledge it was admissible, as in eighteenth-century dynamics and chemistry and in nineteenth-century bacteriology, it could be used to put in order whole fields of genuine but chaotic knowledge. Elsewhere it tended to degenerate into arid commonplaces and false simplifications.

Since 1945, the relationship has been in unresolved crisis: on the one hand, the consequences of science to the social order are too important to be relegated to the sidelines. On the other, the political establishment of all persuasions prefer to "Keep scientists on tap, not on top." Insufferable as this doctrine is to all scientists, they must take care to ask whether they can achieve a greater influence on policy without also invoking more political control of the conduct of science. My essay picks up from Descartes' dilemma. After reviewing the diverse roles of scientists in contemporary society, I will return to some prescriptions about their overarching responsibilities—about the social function of the scientist. In my view, to tell the truth is the categorical imperative.

In contemporary society, the scientist is the one who discovers. We should complicate our definition of discovery, usually given as the uncovering of new knowledge; hidden here is the premise that all "old knowledge" is visible and understood. Furthermore, what is discovered must be important, it must meet some canonical criteria of significance; we look implicitly for an extension of our "understanding" of the natural world. This embraces experimental facts, but it also embraces, just as importantly, theoretical insight and the recording, communication, persuasion and dialectic of those insights. All this implies a community of scientists. Without such engagement in that community, without a forum for insistence on an organized, skeptical criticism of conceptually novel claims, factual discovery would be useless for further increments, would in fact be totally sterile. Science is inherently a social enterprise; an important social function of science is to design and manage its own organization so as to optimize the creative possibilities of its practitioners, and at the same time generate the fruits that justify the ever more costly social investment needed for science to continue. As Bernal repeatedly insisted, and indeed it rings true to this day, most political establishments are relatively unsophisticated in their understanding of the essentiality, difficulties, and inevitably long time scales of basic scientific research. They tend to be captivated by nicely encapsulated, albeit sometimes very costly, projects whose goals appear to be well-defined—all this at the expense of maintaining an alert community able to create and capitalize on the most important—which are always the unexpected—discoveries.

My first assertion is that the preeminent social responsibility of the scientist is the integrity of science itself: to engage in discovery to its farthest reaches as a personal goal, to be part of the community of discussion and criticism, to maintain the ethics of truth-telling, to use no standards other than those of scientific accomplishment in the selection and operation of the managers and gatekeepers of science. To satisfy these responsibilities goes beyond being the most efficient technician in the elicitation of scientific fact, which is the orientation of today's highly specialized disciplinary training. It requires relentless criticism of others' ideas, and equity and compassion in dealing with their claims for personal standing. It may require a broader study of the reaches of science, so as to explore their interconnections, than is achievable in school; and likewise an attentiveness to history, to an understanding of what is known, that may be a momentary distraction from today's new experiment.

This ideal is not always congruent with the interests of the organization, the corporation or the state. The truth is not always the superordinate goal of political affairs, self-deception being even more prevalent than malice. But if scientists ever compromise themselves on this principle, Nature will be no more forgiving than will be a society which has nowhere else to turn for objective analysis of technically convoluted affairs.

Descartes' compromise was negotiated under *force majeure*; it was motivated by the commitment to saving science's integrity within the sphere in which it could authentically operate. His patience paid off: all the theological fuss about heliocentrism and evolution has hardly impaired the claim of contemporary dogmas to spiritual authority. They have had to acknowledge that their primitive pioneers were overzealous in looking to descriptions of the natural world as having any bearing on the eschatological province. Bruno and Vavilov were both victims of crude fundamentalisms that no one defends today.

In the present era, scientists are often called upon, and some volunteer as well, to give advice to society on a multitude of questions requiring scientific expertise. Many of these fall in the category of risk-cost-benefit analyses; the greatest frustration of the scientific expert is in dealing with expectations of perfect safety or zero pollution.[3] At the next

---

[3] In contradiction to the implications of a scientific tenet such as Avogadro's number, 6×10^{23} molecules per gram-mole, that assures us that every breath we inhale contains at least one particle of Nefertiti's perfume.

step of the analysis, it may be equally frustrating to be driven to conclusions when the evidential basis remains tantalizingly fragile. However, the scientist has the ability *and the responsibility* to bring to the analysis the same attention to objective fact, and its delineation from value inclination, as inherent in an experiment. It is impossible to free oneself from bias, but the exercise of scientific judgment within the discipline of the peer group can go far to identify which are the value-oriented and which the scientific underpinnings of the tradeoff analysis.

A byproduct of playing a key role in major social decisions is the double-edged scalpel of political power. Many scientists seek more influence in the political process, partly out of a conviction of what the scientific mind can bring to it, partly for the usual human motives of ambition and quest for power and prestige. I have no doubt that governments could be vastly improved by changing the proportion of scientists to lawyers in its legislatures and at the top reaches of the executive. (So might other organizations, e.g. corporations and even universities.) The danger is the inversion of the process: can scientists live at the court of the prince? Can they gain more political power and prestige without the intrusion of political criteria for advancement within the scientific community? Can they achieve their fair share of affluence without being corrupted? Where else can society turn for untarnished advice on scientific matters that may have immense political and economic consequences?

Finally there is the unbidden advice, the foresight about future extrapolations for which early warning may have inestimable social value. It is said that "prediction is difficult, especially about the future." However, scientists are better experienced than most prophets in articulating predictions as hypotheses; the ability to make confirmable predictions is the core of experimental science. That art, together with a technically complex understanding of matters pertaining to the environment, to human biology, to weapons effects, to technological capabilities of different groups or countries over time, is indispensable in helping a society foresee the long term consequences of its policies in all those spheres. Many scientific advances in this century—nuclear fission is the prototype—have elicited well-founded anxieties about the compatibility of quarreling national sovereignties with the survival of human culture. We are so far from a feasible world model of supranational control of such enormous powers of destruction that scientists today have a special responsibility to assist in the design of the interim arrangements of international accommodation to domesticate such powers. I say scientists, for it is unlikely that other vocations have offered a comparably precise realism about the destructive power at stake or the possibilities of its containment.

Some say that scientists in a given country should simply refrain from conducting science that could have such fruits. How futile that is!

On the one hand, who could have foreseen that studying atomic structure, teasing out the neurons, could so quickly result in weapons? We would have to suspend all science in order to give that assurance. On the other hand, that abjuration might offer some self-satisfaction to the individual scientist, but it can hardly alter natural fact. Instead, it merely assures that the technological breakthroughs will be the monopoly of the most unscrupulous. Even with their limited prophetic vision, nevertheless, scientists are uniquely situated to extrapolate the future possible early warning to what "society" must do to reap the most benefits and risk the least harm. Today's world, divided north-south as well as east-west, offers many impediments to constructive responses to global threats, be they from natural, social or technological sources. All the more reason for the utmost clarity of foresight. Those foresights, together with the inherent supra-national character of scientific advance, have made the scientific profession uniquely motivated and practiced in sustained international concern and dialogue. This is already enough to alarm sovereign states, which have sought to humiliate an Oppenheimer[4] and keep a Sakharov[5] in internal exile. In the past, a country that constrained scientific freedom could do great injury to its own development, as we know from the examples of the geneticist Vavilov in the Soviet Union, and from the Jewish scientists exiled from Germany and Italy before World War II. Today, there is an even broader stake. I wrote these lines during an exhilarating turnaround of East-West perspectives on nuclear arms control:

> For the first time in decades, we foresee the possibility of reversing the accumulation of the most destructive weapons. The broad range of political conflict aside, the fear generated by these weapons has achieved a life of its own in sustaining security anxieties. We see bold proposals, and new approaches to verification including on-site inspection, that were unimaginable a few years ago.

In a short time, "Glasnost" promised to reopen an unprecedented scope of individual expression in the Soviet Union. We could therefore be newly optimistic, and should have been correspondingly insistent, about the development of a framework for East-West confidence leading to still more comprehensive measures. Scientists could play a special role as monitors of sovereign compliance with international order. They have

---

[4] Michelmore, P. The Swift Years: The Robert Oppenheimer Story. NY: Dodd, Mead, 1969. And Stern, P. The Oppenheimer Case: Security on Trial. NY: Harper and Rowe, 1969.

[5] Sakharov, A.D. My Country and the World. NY: Alfred A. Knopf, 1975.

the skills, they have the motivation; it remains for them to receive and sustain the freedom.

In the long run, and when the gravest security interests are at stake—as applies with the most substantial arms reductions—self-inspection and self-monitoring must be a centerpiece of verification and compliance. The investigative press, the Congress, and the concerned scientists play that role in the U.S. to a degree that only Glasnost would permit us to imagine in communist countries. In all countries, still more robust legal guarantees of freedom of access to information and of expression, of assembly and of movement are necessary. The traditions of truth, international communication, supranational concern, and personal courage have marked many notable scientists as trusted guardians of shared values. They are often nuisances to the established order, sometimes to the tranquility of their own fraternity. But they may be the keystone of the phasedown of arms and of international hostility, which is the precondition of survival for any stable social order. All governments share that goal. The commitment of every government is indispensable to assure the freedom of expression of its own scientists, to make them credible as tellers of the truth. Scientists, too, must make themselves worthy of that confidence.

Epilogue

An earlier version of this article was composed for the Bernal centenary and originally completed in March 1988. At that time, the Akademie der Wissenschaften der DDR (East German Academy of Sciences) was well ensconced behind the Berlin Wall. I wondered how far they would be able to go in accepting a contribution that was dissonant of the prevailing socialist ideology (to which J.D. Bernal had, in principle, subscribed during his lifetime). I did not want to offer gratuitous insult, but I could not ignore the historical repression of genetic theory associated with the name of Trofim Lysenko,[6] Nor the victimization of Andrei Sakharov. Nor could I fairly exclude the history of J.R. Oppenheimer as a parallel example in the United States.

My primary prescription for the social role of the scientist was to do authentic science: to sustain the bright flame of objective inquiry as process and product in a world tainted with special and subjectivist interests. This is not only to ensure that public policy and people's lives are governed by a realistic appraisal of the world that appealed to an externally validated, asymptomatically achievable truth. That might even be a prototype for the conduct of our political and social affairs.

Sakharov and Oppenheimer were examples of scientists who, from that platform, also perceived much broader responsibilities for

---

[6] Sayfer, V. Lysenko and the Tragedy of Soviet Science. Rutgers University Press, 1994.

scientists in the public world. Each had been the leading architect of his country's nuclear weapon; each understood all too well the consequences of a belief that such a weapon could ever again be used in any role other than mutual deterrence. Certainly, each went beyond the conduct of science in his tireless efforts to teach that lesson, but neither lost his credibility of adherence to the scientific method in the structuring of his arguments. They are heroic role models, but their intellectual power and historic influence went far beyond the imaginable horizons of most scientists today; we now seem to restrict ourselves to more modest aspirations. So I return again to "do good science!" and sustain the credibility of that profession in public counsels.

Before the Bernal centenary book reached print, the Berlin Wall had tumbled, and I came to realize that the academicians may even have welcomed a level of dissent about communist orthodoxy they may have been too intimidated to express on their own. I have never visited Berlin, but even in Moscow I had encountered many scientific colleagues who regarded the orthodox doctrine as a joke: Why are philosophers' brains so expensive?—Because they are so hard to find. "Philosophers" was the code word for the ideologues. The sustained communication among scientists surely did mitigate the Cold War. During the early days of *perestroika* it also allowed for some return of mutual confidence that Americans were not about to exploit reform in the Soviet system to endanger the physical security of citizens behind the iron curtain. Scientists continue to play such roles, in trying to broker more sensible arrangements for economic recovery in Eastern Europe, and in programs to help finance conversion from nuclear and other weapons to more peaceful pursuits.

RACE, GENES AND HUMAN ORIGINS:
HOW GENETICALLY DIVERSE ARE WE?
Kenneth K. Kidd

Alexander Pope may have used language considered sexist today when he wrote, "Know then thyself, presume not God to scan; the proper study of mankind is man."[1] However, it is clearly the belief of many scientists today that he was correct and that we ourselves constitute a proper subject for study. And there is no more fundamental aspect of humankind than the human genome. Recently we have seen a great many articles in major newspapers and magazines about cloning genes; people in all walks of life are aware of the human genome. One major component of the Genome Project is the organized effort to obtain a representative DNA sequence for *Homo sapiens*, to know the identities of all the nucleotide bases ("letters") in the 3.2×10^9 individual positions in the human DNA sequence, in their proper order. Among the numerous motivations for studying the DNA sequence is to identify the genes that, when defective, cause human diseases.

When we think beyond the genome of an individual to the genome of a species, we soon realize that the latter is not a single representative DNA sequence but also includes the natural variation in DNA sequence among individuals, the "collection" of all the millions of varied DNA sequences. Sequence variation is normal and natural; this variation makes every individual different from every other individual (except identical twins). Normal variation is obvious; it is what we use to recognize individuals—differences in height, shape of the nose, color of the skin, sound of the voice, etc. It is the basic DNA blueprint that makes us humans, and it is the variation in the DNA sequences that generates so much of the biological difference from one person to another. Consequently, if we are going to understand the human genome, we have to understand the variation in the DNA sequence present in *Homo sapiens*, which causes those and myriad other obvious as well as subtle differences. Humans are not unique in the fact of variation; genetic variation is a hallmark of all successful mammalian species. With modern molecular genetic techniques it is now possible to identify and study DNA sequence variation among individuals, but it is important that the studies be inclusive of all *Homo sapiens* or, if more limited, be clearly understood to apply to only some humans.

As noted above, much of the obvious normal variation among individuals is ultimately genetic in origin. It is not racist, on one hand, to study that variation or to describe people according to the geographic

---

[1] Alexander Pope, <u>An Essay on Man</u> (1733-1734), Epistle II, 1.1.

origins of their biological antecedents. It would be racist, on the other hand, to make social value judgments and qualitative classifications of entire populations based on genetic data. Indeed, we contend that *the genetic data on human populations from around the world provide some of the strongest proof against race-based social classifications and stereotypes.* One purpose of this overview is to present enough of the developing data on human genome diversity to support that assertion.

The conceptual relationship between basic human biology and the study of human populations is easily given using two quotations. The original and most famous is by Theodosius Dobzhansky: "Nothing in biology makes sense except in the light of evolution."[2] Gabriel Dover has added a corollary: "Nothing in evolution makes sense except in the light of populations."[3] Understanding the genetic variation within and among populations of humans is a major component of understanding the biology of *Homo sapiens.*

DNA studies of humans, mostly done within the past decade, have begun to provide new insights into the evolution of the human genome. Much of the biological history of our species is written in our DNA sequences. The problem we have is knowing how to read that historical message. If we can read that message, we can begin to address questions such as: Where did modern humans originate? How did modern humans get to all parts of the world? and when? Toward that goal, the scientific developments in the late 1970s and the 1980s were extraordinarily important. Scientists began applying modern molecular biology techniques to studies of human DNA and discovered tremendous amounts of variation detectable directly in the DNA. The fodder for human population genetics prior to the 1980s had been blood groups and serum proteins, now called the "classical" genetic markers. The large numbers of DNA polymorphisms (genes with many arrangements of the four constituent bases A,T,C,G) provided far more variation than geneticists had ever before been able to study with those classical markers and opened completely new areas of research on humans. These DNA polymorphisms have become one of our most important tools for studying human evolutionary history. Studies of the amounts, types, global distribution, and organization of the variations are all ways to "read" the genetic message on human history.

How is genetic variation distributed within and among populations?

In 1972 Richard Lewontin analyzed data from populations around the world on blood groups and other normal genetic variation detected in serum and cells of blood. Data on blood groups were extensive because of their importance in matching donors with recipients

---

[2] 1972.
[3] Unpublished.

for blood transfusions. He attempted to quantify how much of the total variation around the world occurred among individuals within a single population, among populations within a region of the world, and among regions of the world. The result was surprising to many people: the vast majority of the variation occurs between individuals within a single population. Only small amounts of additional variation were contributed by the comparisons among populations within a region and among regions. Many investigators would have considered those comparisons among regions to be comparisons among "races." Nevertheless, very little biological and genetic difference was revealed among regions. A more comprehensive assembly of data on the same blood groups, plus other differences in proteins and antigens, has been analyzed in a variety of ways by Cavalli-Sforza, Menozzi, and Piazza,[4] with a similar conclusion. These studies provided an indirect appraisal of variation in the DNA; however, although hundreds of populations were studied, only a few genes were studied, and therefore only a very small fraction of the DNA was being assessed. Moreover, the variation was known to be extensive between individuals before it was ever studied on populations in different parts of the world; this may have given an inflated estimate of variation between individuals and an underestimate of variation between regions of the world. As it turns out, recent studies examining DNA sequence variation directly, at many positions, have provided strong support for the conclusions of Lewontin[5] and Cavalli-Sforza, et al.[6] that most variation occurs *within* populations.

DNA studies have bolstered our understanding of the distribution of genome variation among individuals and around the world. There are many types of DNA sequence variation, and they need to be considered separately. One common polymorphism involves repeats of portions of the DNA sequence, commonly called short tandemly repeating sequences (STRPs). Many STRP loci have been defined,[7] and each of those documented in the scientific literature has several different common alleles.[8] Individuals of European ancestry will have two different alleles at a given STRP locus more than 70% of the time. This makes these STRPs especially valuable as markers for following how segments of all chromosomes are transmitted from parent to child to grandchild, through families; they are used extensively in studies to map and identify genes that cause inherited diseases.

We have studied 94 STRP loci distributed across several human chromosomes. These STRPs were all selected to be highly informative in

---

[4] 1994.

[5] 1972.

[6] 1994.

[7] E.g., Dib et al., 1996.

[8] Different forms of the same gene, producing different end results.

individuals of European ancestry (largely because those were the samples readily available to the North American and European laboratories conducting these studies), using heterozygosity as a measure of how informative these marker loci are. The question was whether they would also be informative for studies of genetic diseases in other populations from other parts of the world. These 94 markers have been studied in 10 populations from around the world—two from Africa, two from Europe, two from East Asia, one from Melanesia, one from northeastern Siberia, one from North America, and one from South America.[9] Figure 1 graphs two aspects of the genetic variation at these loci in the 10 different populations. The average percent of individuals heterozygous at one of these loci varies from a high of 81% in one of the African populations to a low of 57% in one of the small South American Indian tribes. This Amazonian population is relatively small and among the most isolated of human populations, but it still has adequate amounts of variation that researchers can use for disease mapping studies. In general, African populations show somewhat greater heterozygosity than Europeans; East Asian, Melanesian, and Siberian populations show somewhat lower heterozygosity than seen in Europeans. This heterozygosity is a clear measure of the degree of genetic variation within each population, and the population with the least variation, the Amazonian one, still has 70% as much variation as the one with the most variation. While some parts of the world have somewhat less variation on average than is seen in Europe or Africa, the fact that the maximum reduction is only 30% makes it clear that no human population is genetically homogenous—high levels of genetic variation are ubiquitous, even in small, isolated populations. Thus, the new DNA data, using many more loci than studies by Lewontin[10] or by Cavalli-Sforza, et al.,[11] are confirming a very important finding of the earlier studies: most of the genetic variation in *Homo sapiens* exists among individuals within each population and that *relatively little additional variation exists between populations*.

Heterozygosity, as shown above, is a measure of both the number of different alleles and how frequent they are in the population. Some alleles occur only in one population or another—we term these "private" alleles. If we consider just the number of different alleles from the perspective of what is unique to each population (or region of the world), we get a similar picture, graphed in Figure 1 as the number of "private" alleles. In this example a private allele is one seen only in that specific population sample, but most such alleles are expected to be present in other populations in the same geographic region. They may

---

[9] Calafell, et al., 1997, and unpublished.
[10] 1972.
[11] 1994.

even be rare alleles in other parts of the world, just not seen in our particular samples. Moreover, most of the "private" alleles here are rare even in the population in which they are seen. Even with these caveats there is a dramatic picture of more alleles unique to a particular African population than alleles unique to any population in any other part of the world. One African population had 63 private alleles at these 94 loci, while the South American population had only one. These "private" alleles are in addition to a large number of alleles shared across multiple populations, but the range among populations illustrates both the greater genetic variation present in sub-Saharan Africa and the fact that most alleles seen in non-African populations also exist in African populations. In other words, alleles in populations outside of Africa tend to be a subset of those present in sub-Saharan Africa. From both perspectives, these DNA studies confirm the insight provided by the earlier studies: *variation among populations is very small* compared to variation within populations. They also emphasize that sub-Saharan African populations have more within-population variation than populations in other parts of the world.

How much genetic information is there in *Homo sapiens*?

From studies of DNA we have gained insight into just how similar human DNA sequences are to those of chimpanzees, bonobos and gorillas. The differences in appearance, physiology, and behavior among the four species are determined by only one to two percent difference in the DNA sequences. Moreover, chimpanzees are as different from gorillas as humans are, just in slightly different ways.

Against this background of the genetic similarity of humans to the other great apes, how much variation can there be among humans? Based on over 20,000 such sites (loci) already identified and the fine-scale studies of limited regions, there are estimated to be hundreds of thousands of variable sites. One recent example of a detailed study is the work by Nickerson, et al.[12] on a single DNA region almost 10,000 basepairs long. They found more than eighty sites that vary and that each individual has, on average, just over 20 differences between the copy inherited from the mother and the one from the father. In general, this is normal variation. Most of those differences do not affect the functioning of the protein produced by this gene, but some rare ones do. This is one of the very few studies on a reasonable scale comparing many individuals across a long stretch of DNA. Extrapolating to all 3×10^9 nucleotides in the human genome, there would be 24 million sites that vary among individuals. A similar extrapolation predicts that two DNA sequences taken at random from two different humans will differ on average by about 0.2 percent. If those two random copies are considered

---

[12] 1998.

to be the genetic contribution from the mother and from the father of an individual, then each individual will, on average, have inherited different information from his/her two parents at about six million positions. Thus, another new insight is just how much genetic variation really exists among humans. The claim that "We are all the same but we are all different" is easy to understand when you compare the *greater than 99 percent identity of DNA sequences among all humans* with the roughly six million positions at which the DNA sequences are expected to differ between two copies within a single individual.

Do populations have unique genes?

The common DNA variants are found in almost all populations around the world. The data summarized above show that most human variation is found within a given population and that there is relatively little variation between populations. The data summarized in Figure 2 derive from a type of variation different from that in Figure 1— biallelic polymorphisms.[13] That is, these are loci at which there are only two alternative forms rather than the many (up to 30) alternative alleles at STRP loci. Figure 2 illustrates the pattern of variation seen at three loci in 19 populations distributed over most parts of the world. The order of the populations is arbitrary other than being grouped by major geographical area, as indicated by the labels across the top of the figure. Each of the three systems graphed has just two alleles, of which one has its frequency plotted.

Several points are obvious from the figure, but the most important one is that for these varying sites both alleles are present in almost all populations, in almost all regions of the world. That is generally true for all common variants. Populations do not differ markedly from one another in what genetic forms are present. The second obvious point is that the frequencies of those variants can differ considerably from one population to another. A third point is much more subtle: populations that are geographically close tend to have similar gene frequencies. The relevant words are *tend* and *similar*; frequencies are rarely identical in any two populations, but it is also uncommon for them to be markedly different in geographically close populations.

This similarity is seen for some of the data in Figure 2. One of the loci illustrated is the PLAT gene (Tissue Plasminogen Activator), which, for our purposes, is a random marker. The PLAT locus has similar allele frequencies in the two African populations, likewise in the two European ones; the European alleles are fairly different from those seen in the African sample. On the other hand, the allele frequencies at the Dopamine D2 Receptor (DRD2) gene, likewise chosen arbitrarily from a very large number of possible markers, are fairly similar among

---

[13] Differences due to variations in two different alleles.

the North American populations and among the South American populations.

These data further support our understanding that most DNA variants exist in most populations, though they occur at different frequencies. Very few common forms are absolutely unique to any one area; the variants more limited in distribution tend to be rare even where they do occur. What variation exists between populations accumulates gradually across large geographic distances; it is, therefore, a significant conclusion that there are no sharp boundaries dividing human groups. This pattern of DNA variation argues strongly that human races, defined as distinct populations with significant biological differences from all others, *do not exist.*

Is variation organized differently in different populations?

Within small segments of DNA, 10,000 to 30,000 basepairs long, the different alleles at nearby sites may not occur at random in all possible combinations on the chromosomes in a population. To understand this, consider just two varying sites, one STRP with 5 alleles and a nearby biallelic marker. If these sites are only 10,000 basepairs apart, the whole segment of DNA containing both sites will almost always be inherited as a single unit from parent to child. A total of 10 different combinations of alleles at the two sites could occur on chromosomes in the population. If the five STRP alleles had equal frequencies of 20% each and both alleles at the biallelic marker had equal frequencies of 50%, the 10 possible chromosomal types would each have a frequency of 10% if all combinations occurred at random chance frequencies. Rarely are stations so simple for humans. Much more likely are unequal frequencies of alleles at each site, and in many populations the combinations occur at non-random frequencies. "Linkage disequilibrium" is the term used to refer to non-randomness in the frequencies of such allelic combinations at closely linked marker sites.

A study of two polymorphic sites around the CD4 gene by Tishkoff, et al.[14] was the first large scale study of linkage disequilibrium in multiple human populations. A total of 12 alleles occurred at an STRP site if all population samples were considered, but only rarely were all 12 seen in a single population sample. The frequencies of the different alleles differed considerably among populations, much as would be expected from the results at many STRP loci, summarized in Figure 1. The sub-Saharan African populations had many different alleles, but only three of the 12 alleles were seen at common frequencies in the non-African populations. The remarkable finding was that the two alleles at the nearby biallelic marker occurred in combination with many different STRP alleles in the sub-Saharan populations, but only with one STRP

---

[14] 1996.

allele in the non-African populations. Moreover, it is the same allelic combinations that are responsible for the non-randomness in all of the non-African populations. Interestingly, populations in Ethiopia and Somalia have intermediate patterns of allelic association and intermediate amounts of non-random allelic association. The measure of the association studies by Tishkoff et al.[15] is graphed in Figure 3.

A similar pattern of very little linkage disequilibrium in sub-Saharan African populations but very strong linkage disequilibrium in all non-African populations has now been seen at other loci as well.[16] At each of the several genetic regions studied in this way, the alleles seen at the individual sites are the same ones in populations outside of Africa as in populations in sub-Saharan Africa (though there are usually more STRP alleles in the sub-Saharan populations). What is most different is the number of combinations of those alleles on chromosomes in the populations. We will deal with the implications of that difference in the next section, but note that even here we see that geographically close populations have very similar combinations. There do seem to be two major regions of the world, sub-Saharan Africa and the rest of the world except Africa, but there are also populations that are intermediate in those parts of Africa that are geographically intermediate. Again, we conclude that it is impossible to draw a line that divides humans into qualitatively different groups.

What does genetic variation tell us about the history of modern humans?

When we focus on the small amount of difference between populations, a distinct pattern of genetic variation among populations emerges, with sub-Saharan African populations having the most genetic variation, European and South West Asian populations less, East Asian populations still less, and Amerindian populations the least. Moreover, each of those broad geographic regions with less variation has a subset of the variation in the adjacent "preceding" region. The successive subsetting of variation is a strong argument for the "Out of Africa" scenario for modern human origins. The difference in amount and pattern of linkage disequilibrium is particularly important because it gives us insight into the ages of populations. We know that normal genetic processes will, though the generations, randomize the allelic combinations on a chromosome. The randomization will occur quickly for genes at opposite ends of a chromosome but very slowly for genes or loci that are very close together molecularly.

We have concluded that modern humans arose in Africa and that one group left Africa and then expanded to fill the rest of the world, with

---

[15] 1996.

[16] Tishkoff, et al., 1998; Kidd, et al., 1998; Zhao, et al., 1997; and unpublished.

successive loss of variation as some first moved out of Africa, then some of their descendants moved from west Asia into central and east Asia, and then some descendants of those early Asians moved into the Americas. We can also infer, from the high levels of disequilibrium and the paucity of "new" variation in populations outside of Africa, that this migration was recent (i.e., about 10,000 years ago). Furthermore, the continuous subsetting of variation with increasing distance from Africa leads to the conclusion that it was the descendants of essentially a single migration out of Africa that expanded eventually to occupy the rest of the world, and also that these modern humans replaced all preexisting humans already present: *Homo erectus* in Asia and *Homo neanderthalensis* in Europe.

Taking into account large numbers of loci such as the three illustrated in Figure 2, it is possible to extract a global pattern that tends to cluster populations geographically and can be interpreted as the consequence of the migrations history just summarized. This is illustrated in Figure 4, a two-dimensional principal components analysis of genetic distances at 15 different genes, many of which differ at multiple sites. The three loci in Figure 2 are part of the data analyzed, but the STRP data illustrated in Figure 1 are not included in this analysis; they give a very similar result.[17] A genetic distance is a calculation of differences in allele frequencies between two populations based on multiple genetic loci. Irrespective of the method used to calculate genetic distances between loci, we see a very clear geographic clustering. It is not surprising that geographically close populations seem to be closely related genetically.

It would be false to interpret the pattern shown in Figure 4 as defining racial groupings for a variety of reasons, the most important being that the selection of populations is not uniform around the world. There are no populations sampled from most of Africa, especially from the east coast of Africa, Ethiopia and Somalia. There are no populations sampled from far eastern Europe or from central and south central Asia. As we add more populations to such an analysis we see that the gaps (or separations between large geographical regions) are filled in, and it becomes much clearer that variation is continuously distributed, not abrupt or discontinuous in its distribution.

The genetic variation among human populations shows a continuous gradation with geographic distance. This variation is interesting and medically relevant—but it is not socially relevant. It is not possible to claim the genetic superiority or inferiority of a population based on its geographical-genetic origins. Furthermore, no definitive boundaries exist among the myriad variations in DNA, so *it cannot even be claimed that dramatically distinct "races" exist among human beings.*

---

[17] Calafell, et al., 1997, and unpublished.

Racist classifications cannot call on genetics research as an ally. What genetics research can and does contribute, through a systematic study of human genetic diversity, is an even greater understanding of how similar we all are, even in our marvelous variation.

Figure 1

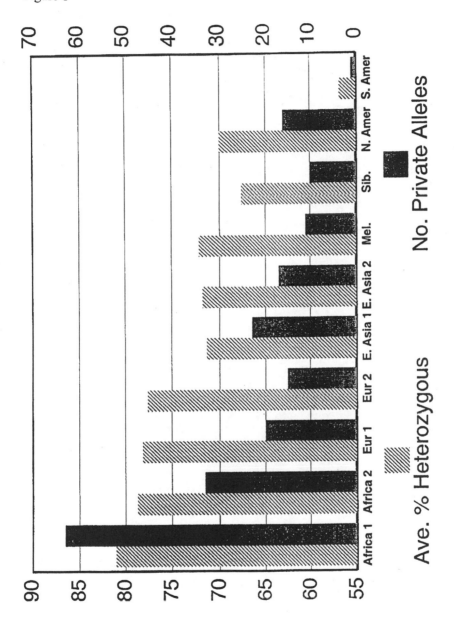

22

Figure 2

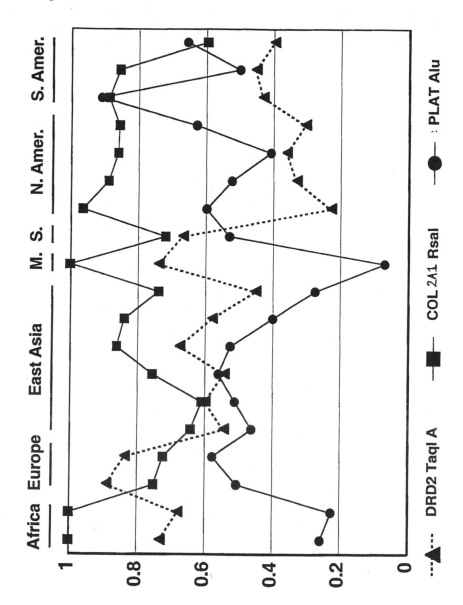

Figure 3

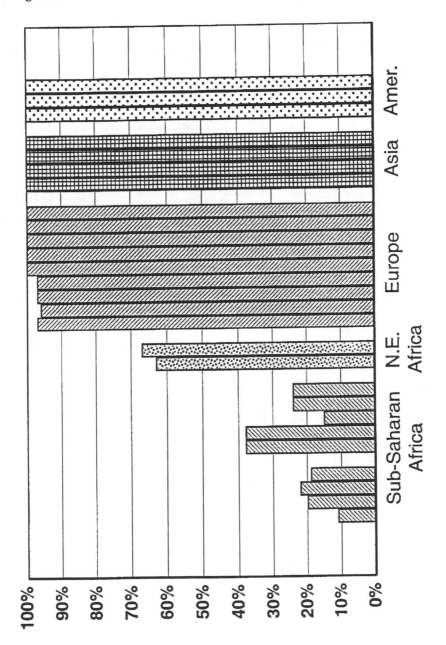

24

Figure 4

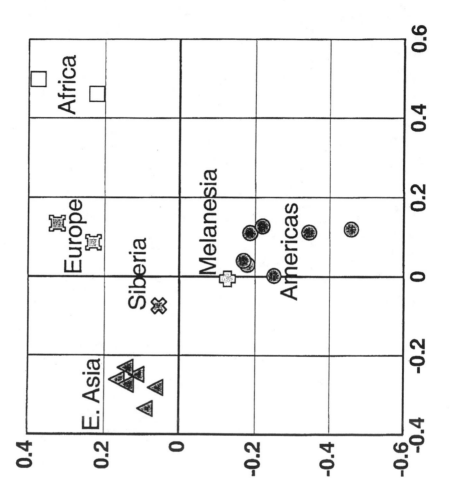

CAN AGRICULTURAL BIOTECHNOLOGY SOLVE WORLD HUNGER PROBLEMS? THE ROLE OF REGULATIONS AND OLIGOPOLIES

Ellen Messer and Nina Dudnik

Introduction

This essay evaluates the promise and practice of agricultural biotechnology (ABT). It considers the first decade of ABT products (1987-1997) and their environmental, social, nutritional and ethical consequences. Since the first plant was transformed by genetic engineering in 1983, the ABT industry has prophesied "win-win" scenarios, with higher-value crops at lower environmental costs, plus products tailored to meet the special nutritional needs of the elderly and other groups.[1] Proponents furthermore promise that ABT, if affordable and accessible for third world countries, would help overcome world hunger by (1) increasing food production, (2) lowering food production and consumption costs, and (3) developing products to meet the special needs of nutritionally deprived groups. ABT's contributions to medical diagnosis and treatment, as well as commercial employment and trade, might also improve the health and incomes of many who might otherwise be hungry.[2]

Critics, by contrast, predict losers as well as winners. They anticipate high risks for environmental and human health, losses in biodiversity, economic displacement and lower incomes for small farmers, and reductions in consumer choice.[3] Moreover, in developing regions, ABT could further compromise the competitive positions of disadvantaged farmers and nations, and could exacerbate the marginality of the rural poor. Critics also caution against promotions which claim that ABT's potential to overcome world hunger takes precedence over exacting environmental-impact and other time-consuming risk evaluations of ABT products, particularly for developing countries.[4]

Both proponents and opponents recognize that; short- and long-term outcomes will depend on how and by whom ABT priorities are set and on particular product development choices. Alternative futures are also influenced by the institutional environment, including government safety regulations, trade rules, business interests and consumer attitudes.[5]

---

[1] Gaull and Goldberg 1991; Biotechnology Industry Organization (BIO) 1998.

[2] Messer and Heywood 1990; Doyle and Persley 1996.

[3] Krimsky and Wrubel 1996; Hodgson 1992; Junne 1992.

[4] E.g. Hobbelink 1991.

[5] Sasson 1988; Messer and Heywood 1990; Hobbelink 1991; Rissler and Mellon 1993.

26

The sections below describe early ABT choices in the US and the bioethics concerns they raise. Taking genetically modified organisms (GMOs) approved for commercial release in the US in mid-1998 as an end point, we examine what transformation techniques, plant species and crop characteristics are emphasized; what business structures and regulatory apparatuses have shaped these early ABT choices; and how commercial, government and other interests raise and respond to questions of environmental impact, food and nutritional safety, social impacts and additional moral concerns.

A brief history of agricultural biotechnology

ABT is defined here as the *in vitro* manipulation of whole plant, cellular or molecular materials for the purpose of improving agricultural plants or processes. In contrast to conventional plant breeding, which transfers pollen between selected individuals in closely related varieties to produce offspring with desirable traits, genetic engineering (GE, the ABT of interest here) utilizes recombinant DNA techniques to transfer genes between species. Through selected gene transfers between species, GE is able to draw on a wider range of genes and generate GMOs with useful phenotypes unachievable by conventional breeding or alternative ABTs (such as mutation selection breeding, anther culture, somaclonal or gametoclonal variation, or protoplast fusion)[6,7]. Cell- and tissue-culture techniques have been available since the 1950s, gene cloning since the 1970s, and plant genetic transformation (GE, recombinant DNA) since the 1980s. In the 1990s scientists significantly routinized the four steps of plant transformation: (1) introduction of foreign (novel) genes (DNA) into plant cells or protoplasts; (2) stable integration of novel DNA into the recipient genetic environment; (3) efficient detection and selection of transformed cells; and (4) regeneration of fertile plants from transformed cells.[8] But there remain many unknowns in each step of the transformation process, plus the fundamental challenge to transfer and control the expression of more genes controlling useful phenotypes.

Additional techno-economic hurdles involve the proprietary interests that surround and limit access to almost all steps in the process from gene transfer to plant multiplication. ABT, in another contrast to

[6] E.g. Cocking 1990.

[7] Mutation selection breeding: Plant is treated with a mutagenic or chemical radiation, then desirable mutants are selected. Anther culture: Excised anthers are cultured, giving rise directly to haploid plants expressing only the male lineage. Somaclonal or gametoclonal variation: Cells grown in culture develop obvious variants. Single cells of the variants may be selected and grown into entire plants. Protoplast fusion: Cell walls are enzymatically removed from cells of two different species. The resulting naked protoplasts can be fused, producing somatic hybrids.

[8] Birch 1997.

conventional plant breeding, is being developed almost entirely in the private sector, where private firms attempt to patent and control the licensing of methods, processes and products. The scientists' search for innovative ways to increase the frequency and predictability of transformation events aims not only to achieve a higher proportion of useful transformants but also to bypass expensive, patented logistical barriers.[9]

Research since 1983, the year the first successful plant transformation was reported, has focused on understanding the mechanisms of gene integration and expression, improving efficiency of transformation, and increasing the number of crop species and traits which can be manipulated. The most widely used technique involves a bacterium, *Agrobacterium tumefaciens*, which induces tumors on its host by transferring a circular piece of DNA, called a plasmid, from its own genome to that of its host. Splicing any desired gene into a plasmid ensures its transfer as well. Thus scientists achieve the movement of a desirable gene from one genetic source to the receptor plant, by means of the bacterium. This process occurs in nature without human involvement, a fact that some suggest establishes GE as a largely "natural intervention." Although this technique was originally limited to broad-leaved (dicot) plants, it has recently been adapted for use with narrow-leaved (monocot) plants as well. Since the cereals belong to the latter group, this advance is particularly helpful in grain production. Less efficient techniques to introduce novel DNA into target organisms involve bombardment by a "gene gun" which "fires" DNA-coated tungsten microprojectiles into the receptor tissue, microinjection by fine needles, and facilitated uptake of DNA by receptor cells. Table 1 describes these methods, some applications, and the regulatory issues they raise.

Developing more crop-specific plant transformation systems is another rapidly advancing area. Worldwide in the 1990s, more than 120 species in 35 families can be transformed, and no species is classified as recalcitrant to gene transfer and regeneration.[10] Some, such as cassava, took longer than anticipated to achieve successful transformation and regeneration systems.[11] As recently as 1992, major monocots such as rice, maize and wheat could not be reliably or routinely transformed. But scientists correctly viewed these as temporary problems which have since been overcome. Although the private sector has been faulted for focusing on major commercial crops and bypassing crops that feed the poor, research in major crops such as maize and rice is accelerating the pace of ABT achievements across species.

---

[9] Ibid.
[10] James and Krattinger 1996.
[11] Thro, Taylor, Raemakers et al. 1998.

The range of crop characteristics that molecular biologists can manipulate is also expanding. Genetic engineers control increasing numbers of genes coding for specific herbicide tolerances, insect and disease resistance, and altered nutrient composition—the relative quantity and quality of carbohydrates, proteins, and fats (usually in seeds). In the US pipeline are varieties of maize and soybean boasting "higher energy" (i.e., more fat, relative to protein and carbohydrates), "healthier" fat, "more balanced" protein, and "designer" starches, each tailored to particular nutritional conditions, industrial processes, or other end uses. [12] Scientists can also manipulate genes that control the timing of reproduction, fertility, sterility and ripening; or that allow plants to express valuable pharmaceutical products or plastics components. In addition, to produce plants with agronomic characteristics such as extended shelf life, scientists have routinized the "anti-sense" approach to controlling gene expression, which introduces into a cell an RNA molecule complementary to the RNA sequence of the target gene, and so blocks translation of the target gene (in this case the one which accelerates ripening) into its protein end product and limits its expression.

Choices of which crops and characteristics to develop are influenced by "what farmers want" (to overcome yield-limiting factors), what is commercially desirable (based on market assessment of what consumers, including processors, want), and what is technologically feasible. GE so far is restricted mainly to "single-gene" or "stacked" single-gene traits, such as selected herbicide tolerance and insect resistance, because these are what scientists so far control. What is commercially desirable depends on ecological constraints, the availability and costs of alternative remediation strategies, and what profits private developers can expect from an ABT advance. As mentioned above, almost all ABT research and development (R&D) is in the private sector, which patents and controls licensing of ABT processes

---

[12] Dupont in mid-1998 was reported to be leading Monsanto (American Home Products) in "second wave" ABT, with more patents for nutritional attributes. It reported "dozens of futuristic ideas on the drawing board. Among them: instructing soybeans to make more of a natural compound that might fight cancer, or making corn that reduces the amount of saturated fat in the eggs of chickens that eat it, Currently, DuPont is contracting with farmers in the Midwest to grow a soybean for making healthier cooking oil. Eventually, DuPont hopes to be able to take orders for a new type of crop from food companies such as Nestle SA or ConAgra, Inc., create it in the laboratory, contract with farmers to grow millions of acres and process it into a food ingredient...." Monsanto planned to use Cargill's system of rural grain elevators to contract with farmers to produce genetically engineered crops and mill them into ingredients "tailored to the specifications of Cargill's customers, including most of the world's biggest food companies." (Kilman and Warren, 1998: B1, 7)

and products. These commercial aspects raise ethical concerns among those who view ABT as desirable and necessary to improve food production at lower environmental costs. They ask difficult, pressing questions, such as *whether ABT will be available to improve crop varieties for disadvantaged peoples and countries, especially in the developing world.* Legal property rights over genes, plant parts and whole plant organisms should also raise *profound ethical issues about human relationships with nature: whether, for example, nature should be commercialized, genes for sale, and any part of the earth's genetic heritage privatized for profit by private interests.*

In the US, ABT is dominated by the private sector, and in particular by a few large firms. These "life-science" conglomerates, which include US-based DuPont and Monsanto, and European-based Novartis and AgrEvo, focus R&D on principal economic crops and their major barriers to production. They increasingly control the seed industries, transformation techniques, genetic constructs, and associated chemicals used with improved seeds. Setting herbicide tolerance as a priority in soybeans, as a case in point, allows Monsanto to sell both its herbicide (Roundup) and seeds genetically engineered to tolerate it.

What is commercially available, however, also depends on what passes through the US regulatory apparatus, whose records provide additional data chronicling ABT's initial 10 years.

Safety regulation

To assure that ABT poses no environmental, food or health risk—nor barriers to trade—all GMOs are subject to international and national safety regulation. Before a product is deregulated for field release and approved for commercialization, its developers must demonstrate stable transmission of the foreign DNA. In addition, they must show that the product poses no environmental hazards, that it will not escape and become a weed, transfer genes to wild relatives that might become noxious pests, or introduce toxins into the environment or end products. They must also demonstrate that it is nutritionally safe, that it does not contain any additional toxins or allergens that could harm consumers, and that it meets nutritional expectations for the generic (conventional) product of that name.

International regulations for the US, Europe, Israel and Japan have been developed by the Organization for Economic Cooperation and Development (OECD), through workshop reports[13] and a series of case studies on "harmonization" or standardization of efforts. These publications offer developmental principles and guidelines from the initial stage of laboratory R&D through field testing, scale-up, large-scale release, and product monitoring. The United Nations Food and

---

[13] OECD 1992, 1993, 1994, 1996.

Agriculture Organization and World Health Organization (FAO-WHO), in their Codex Guidelines and Standards,[14] present international standards for countries participating in international trade through the World Trade Organization or General Agreement on Tariffs and Trade (GATT).

In the US, the National Research Council in 1984 recommended a safety assessment framework in which ABT crops must win approval of three regulatory agencies: the United Stated Department of Agriculture (USDA), Food and Drug Administration (FDA), and Environmental Protection Agency (EPA). USDA's Animal and Plant Health Inspection Service (APHIS) sets the experimental protocol for field release (field testing) by which GMOs are evaluated as "safe." Applications for "permits" are the first and most stringent procedure. Later applications for similar GMOs are approved by more streamlined "notification" and "petition" procedures. FDA establishes the principles and protocol by which the end product is judged nutritionally "safe," and found not to introduce any new toxins, allergens or anti-nutritive factors. EPA must certify that neither the plant nor associated products introduce harm into the environment. Any introduced toxin must prove non-persistent and harmless to non-target species. In addition, the process of producing transformed plants must introduce no hazardous products, e.g. laboratory wastes, into the environment.

Both international and US assessments are based on the principle of familiarity of "substantial equivalence." GMOs are judged to be "safe" if they can be shown to be "equivalent" to a known, conventional organism, usually a conventional variety of the same crop. Cumulative data on the phenotypic expression of the novel genetic trait(s), and their physical and biological interactions in known environments, are used to establish "known" classes of organisms as safe. Product developers must then demonstrate that their GMOs are "substantially equivalent" to these conventional organisms and pose no new threat. Novel foods which are not substantially equivalent to any existing plant food are evaluated by characterizing the host and donor species, the vector and transgenes used, and the role the crop will play in the ecosystem or the food in the diet. International agencies are developing databases of the information required to evaluate risks of field release or to perform substantial equivalence comparisons for food safety evaluation.

Against this regulatory background, private commercial firms have been trying to hurry GMOs to market. USDA APHIS numbers and graphics indicate their progress. Figure 1 displays the rapidly increasing numbers of GMO field releases over 10 years: from five in 1987 to 747 in 1997. And 86% of those in 1997 were approved under notification, a less onerous procedure. Figure 2 indicates the multiplying numbers of

---

[14] 1997.

field test sites, which grew from five in 1987 to almost 4000 in 1995 and 1997 as product developers began to test GMOs in multiple environments. Over this period, APHIS approved field releases in 44 different plant species. Maize was the most frequent crop tested, with 1420 field releases between 1987 and November 1997. Tomato, potato and soybean followed, each with 300+ field releases, and cotton with 237. 26 species had fewer than five releases.

Field releases involved nine official categories of phenotypes (plant characteristics). Between 1987 and the end of November 1997, APHIS reported issuing a total of almost 3700 permits and acknowledged notifications. The largest number (29.0%) were for the "herbicide tolerance" phenotype. Insect resistance (24.0%) and product quality (21.2%) were the other largest categories. Viral and fungal resistance accounted for 9.8% and 4.2%, respectively. Marker genes, bacterial resistance and nematode resistance were the categories least studied; grouped together they comprise only 7.4% of total releases. Those releases involving agronomic properties such as environmental stress tolerance accounted for only 4.4% of the total. These data are summarized in Figure 3.

APHIS data also reveal that the majority of GMOs are being developed in the private sector by a few large firms. This is shown clearly in the case of maize, the crop most frequently modified. Of all permits, notifications and petitions for maize approved by APHIS between 1993 and the end of 1997,[15] only three were submitted by (USDA) Agricultural Research Stations, plus 22 by eight different universities. The remaining applications, a total of over 1800, were completed by 36 private companies.[16] More than 1300 of these were submitted by only five companies—DeKalb Genetics, Monsanto, DuPont, Pioneer Hi-Bred, and Northrup-King (Novartis). Furthermore, by mid-1998 DeKalb was wholly owned by Monsanto, and DuPont owned a substantial stake in Pioneer Hi-Bred.

Approvals and commercial releases

Since 1994, the year Calgene's Flav'r Sav'r (MacGregor) tomato became the first GMO authorized for commercialization, the prevalence of GMOs in US agriculture has soared from no GMOs to a full 20% of maize, 40% of soybeans, and 50% of cotton. Monsanto, the US leader in early ABT, released its first commercial varieties—Bt-modified[17] and herbicide-resistant cotton—only in 1996. Just two years later, according

---

[15] The database was accessed on December 2, 1997.

[16] The numbers include documents that were judged incomplete, denied, pending or withdrawn.

[17] *Bacillus thuringiensis*, a bacterium producing an insecticide. The gene for this trait has been transferred to several crop plants.

to its 1997 Annual Report, Monsanto was selling 11 genetically-engineered products, projected to cover 50 million acres. Table 2 summarizes the status of GMOs approved by USDA and FDA and authorized for commercial release by mid-1998. Of the initial 34 approved varieties, 8 were maize, followed by 6 tomato, 5 soybean, 5 cotton, 4 canola,[18] 3 potato, 2 squash, and 1 chicory. This set includes three of the most important economic crops in the US—maize, soy, and cotton—plus canola, an important economic crop in Canada, where genetically-engineered varieties were initially registered and sold. Fruits and vegetables with the exception of tomato (a "model" transgenic crop) lag farther behind. Their lag may reflect their lower value (in returns) relative to development costs for the particular crops, techno-agronomic constraints, or both. Certainly the early entries, Calgene's Flav'r Sav'r tomato and Upjohn/Asgrow's virus-resistant squashes, have not been market-pleasers because the tomato bruises easily and requires enhanced resistance to insects and diseases, and the squashes are attacked by various viruses to which they are not yet resistant. However, new varieties of squash, peppers, and other vegetables carrying genes that control ripening are reported in the pipeline.[19]

The dominant crop characteristics are insect resistance based on Bt-toxin, and herbicide tolerance based on genes that block damage by proprietary herbicides. In a few cases both qualities are "stacked" in combination. Five companies—Calgene, DNAP, Zeneca-Petoseed, Monsanto, and Agritope—have tomatoes containing genes for altered fruit ripening. The data also show that ABT in each crop is dominated by a few private firms. For the major field crops, chemical-seed conglomerates have been able to create synergies among research and development capacities in genetic engineering, chemical (protein) engineering, and pharmaceuticals. These synergies add economies of scale through vertical integration and shared research and processes. In addition, only big firms will have deep enough pockets to register initial products, which can then be multiplied through "extension." As these firms grow through mergers or joint ventures they also eliminate the time and expense of licensing, competition and legal battles over patents and proprietary technologies. Acquiring seed companies, they anticipate from a market and extension perspective that GMO-seed sales will be more reliable where farmers can continue to deal with their accustomed seed vendors rather than a new chemical-seed supplier. In the future, farm commodity markets envision themselves supplying differentiated seed products to niche markets that, in the case of herbicide-tolerant crops, also include sales of their chemicals.

---

[18] Also known as oilseed rape.
[19] BIO 1998.

Are there any surprises?

Considering the explosive growth of GMOs over the past five years, with even more rapid growth predicted, it is surprising how few GMOs are yet in the marketplace. Wheat and rice are already in the pipeline, as are a variety of high-value oilseeds genetically engineered for "high energy," or special fats or more nutritious protein composition; and rot-resistant fruits and vegetables. Multiple genes for product-specific herbicide tolerances, and insect and viral resistances, are also close to market approval in diverse crops. But genes and gene products which might be used to introduce "natural" biological soil enhancement through rhizobia, or enhanced productivity based on genetic tinkering with the photosynthetic process, are distant possibilities.

Another surprise is the escalating scale and pace of mergers and acquisitions. In the 1990s, large chemical companies—reinventing themselves as "life sciences" firms—have built internal biotechnology capacities by incorporating wholly owned subsidiaries (start-up companies or those specializing in certain crops; e.g. Monsanto acquired Calgene in 1997), by mergers (Ciba-Geigy and Sandoz merged to form Novartis in 1996), and by joint ventures (Hoechst-Schering launched AgroEvo as a joint venture in 1994). What is not clear is what impact mergers, acquisitions and joint ventures will have on ABT strategies in developing countries, especially where there have been additional joint ventures by private firms and foundations or public institutions. Joint ventures between private ABT firms and public institutions are not philanthropic. The private firms gain the advantage of seeing their genes tested in many more crops and crop varieties, by the best third world scientists, under strict oversight. They also reserve the right to use all findings and products in crops and markets over which they retain control. Furthermore, international agricultural research centers such as the Center for the Improvement of Maize and Wheat (CIMMYT) are advancing ABTs that should prove useful to the companies, and some prior history in negotiated partnerships serves both sides. [20]

Are the products safe? Have they been adequately safety-regulated?

With reference to the principle of "similarity," numerous new transformants of common cultivars such as maize, and new cultivars that incorporate known genes by known vector processes, should multiply the numbers of GMOs in the marketplace over the next ten years. But this will happen only if the public judges them to be nutritionally and environmentally safe, with anticipated benefits outweighing risks. Europeans and Americans at least in this initial period appear to differ on what they expect to be the risks and benefits of GMOs, and have shown greater reluctance to embrace or even permit their testing. Early ABT

---

[20] This is the case for apomixis. See also Messer 1992.

assessments describe four kinds of risks in particular—environmental, ecological, social and nutritional:

(1) impacts on the physical environment, which might also damage human or animal health;

(2) impacts on the biological environment, because early ABT choices could reduce biodiversity in agricultural fields and farmers' choices in varietal planting materials;

(3) impacts on agrarian structure, because ABT could reduce the viability of small farms, the independence and incomes of small farmers, and the markets for products from small-scale agricultural industry;

(4) and impacts on the nutrient quality of genetically engineered products that could also remove a dimension of consumer choice (i.e. not to eat GMOs).

Other serious questions of moral and social propriety surround GMO regulation: everyone from clergy to royalty question the wisdom of tampering with nature and favoring society's commercial needs over the natural order by consuming GMOs; many assert that GMOs tamper dangerously with humans' moral relationship to the natural and superhuman world.

Impacts on the physical environment

Herbicide-tolerant crops are designed to increase the use of certain "safer," more "environmentally friendly" chemical weed-killers. Such herbicides are intended to be used in lower doses, administered by safer (topical) applications, and supposed to be non-persistent in the environment. They replace more toxic chemical herbicides that do persist, polluting groundwaters and soils. Nevertheless, environmental protection interests fear that an emphasis on herbicide-tolerance could result in heightened, not reduced, use of chemical herbicides and their dangerous persistence in the environment. Early evidence supports this contention, although careful and strategic use of the new generation of herbicides, precisely applied, should prove safer than the older herbicides they are replacing. In addition, there is a recognized need for more information—and possible additional regulation—relative to scale effects, which will likely be met only by tests of time or persistent pressure by environmentalists.

Environmental protection and sustainable farmer interests also worry that not all herbicides designed to be used with GMOs are adequately effective or benign. Despite APHIS (USDA) and EPA approvals, evaluations of early commercial releases suggests that these herbicides (glyphosphate, glusofinate, pyrithobac sodium, etc.) are not yet adequately controlled. Bromoxynil, used on cotton, was withdrawn in

1997-1998 pending EPA review; and Monsanto consequently withdrew its bromoxynil-resistant cotton. "Staple" (pyrithobac sodium) appeared to damage non-tolerant crops such as onions and sugar beets grown in rotation with tolerant cotton even after two seasons, a finding which suggests that the herbicide is more persistent than expected. It was also tolerated by weeds in the goosefoot family, which still have to be removed by other chemicals or crop-protection methods. [21]

Agricultural experts and farmers also worry that sowing herbicide-tolerant Roundup-Ready (Monsanto) or Liberty-Link (AgroEvo) varieties of soybeans and maize in rotation could lead to herbicide persistence and possible co-evolution of weed-resistance in affected environments. Herbicide-tolerant crops initially are a good way to rescue a weed-infested field, and their novelty value has created high demand for overall adoption. But usage still needs to evolve to a point where herbicide-tolerant crops become one component tool of an overall management strategy, and do not threaten biodiversity.

Impacts on the biological environment and biodiversity

1998 patterns of GMO usage raise the risk that early adopters will saturate areas with a few elite varieties, decrease crop biodiversity, and thereby increase crop vulnerability to pathogens and to crop pests that will co-evolve. Reduction in biodiversity of field crops, perhaps only a temporary bottleneck, is a principal concern. So far, it is not evident that introduced genes from herbicide-tolerant varieties of maize, soybean and cotton escape into locally-adapted field varieties. It is still very early for Monsanto and other large firms racing to market with herbicide-tolerant and insect-resistant products to be distributing diverse numbers of elite lines with these traits. Evidence of early adoption does suggest, however, that maize, soybean and cotton farmers want seeds that produce integrated management strategies geared to avoid weed and insect damage, both major production constraints. Seed companies, working as part or partners of the chemical giants that control the genes, appear to

---

[21] Farmer advisories cautioned that the herbicide, though effective against nightshades, pigweed, cocklebur, and other broadleaf weeds, only partially controlled morning glory and did not reduce weeds in the goosefoot family, so that additional chemicals or other methods would be necessary in crop protection. The herbicide had to be applied before the nightshades reached the 5- to 6-leaf stage and only in an 8 to 10 inch band over the top of the cotton row. Only this manner of application could lessen cost and residual herbal activity on rotational crops. Research indicated that staple residues could injure other crops (alfalfa, barley, maize, onions, sugar beets, tomatoes, wheat) sometimes grown in rotation with cotton; even after 2 years, yields of onions and sugar beets were reduced. Acala but not Pima varieties of cotton tolerated the herbicide. All this suggests that much more work needs to be done before herbicides can be deemed "non-persistent" in the environment and "safe" (Vargas, R. and Wright, S. 1996 "Staple" registered for the 1996 season. California Cotton Review 39).

sense farmer demand, in addition to their own anticipated higher profits from premium genetically modified seeds. Together, seed buyers and sellers, all wanting to sow the same limited number of varieties, create potential vulnerability to common susceptibility, a kind of tragedy of the commons. Thus, a limited genetic diversity confers the danger of widespread crop annihilation, for instance by a single pathogen to which the GE crops are not immune.

Concern also surrounds the possibility of inadvertently creating "superpests" as a result of cultivars passing genes to wild weedy relatives.[22] Early choices in ABT have sidestepped these concerns, although the United Nations Conference on Environment and Development requires signatories to:

> establish or maintain means to regulate, manage or control the risks associated with the use and release of living modified organisms resulting from biotechnology *which are likely to have adverse environmental impacts that could affect the conservation and sustainable use of biological diversity, taking also into account the risks to human health.*[23]

The Conference also calls for a protocol that will set out safety procedures for:

> safe transfer, handling, and use of any living organism resulting from biotechnology *that may have adverse effects on the conservation and sustainable use of biological diversity.*[24]

In counterpoint, World Bank consultants Doyle and Persley[25] assert that the risks of reduced biodiversity resulting from first generation ABT products are "infinitesimal" compared with the risk of reduction through native habitat loss. They also allege that the risk of novel gene transfer to wild or weedy relatives that might "contaminate" pristine gene pools is very small because there are natural selective pressures against gene flow from cultivars to wild relatives. They reason that "The vast majority of new crop cultivars being produced with techniques of modern biotechnology have been modified to sustain or increase yields" through increased resistance to pests or diseases or increasing ability to compete, as by eliminating weeds.

---

[22] Mikkelson, Anderson, and Jorgensen, 1996.
[23] Agenda 21, Convention on Biodiversity, Article 8g. Emphasis added.
[24] Ibid., Article 19 (3). Emphasis added.
[25] Doyle and Persley, 1996:9.

> In the vast majority of cases ... the pests or diseases
> detrimental to agricultural yields are not the limiting
> environmental constraints on the wild relative receptive to
> outbreeding ... . Experience shows that selection pressures
> found in nature do not favor such gene flow from modified
> crops to wild relatives.[26]

They assert that the risk of gene transfer to weedy relatives in the US is especially low, because so many crops are introduced and so do not face competition from wild or weedy relatives.

Such hyperbole and reference to "relative" risk notwithstanding, more work is needed to base these assertions on scientific assessment. The fact of gene flow between sorghum and Johnson grass (a weed), and between cultivated canola and related weedy brassicas, suggests that the "gene escape" problem is real. This conclusion is reinforced by efforts to eliminate the problem, as by the genetic engineering of plastids, which would put novel genes into chloroplasts rather than into reproductive cells and prevent dissemination of novel genes through pollen.[27] Critics argue that such transformations are difficult, however. Even where successful, they might not prevent weeds from hybridizing with crop relatives by other mechanisms, as where transgenes in the cultivated crop are maternally transmitted, or where crop species have paternal or bilateral inheritance of plastids.

This being the case, environmental protection interests contend that the most important tool to prevent gene escape is vigilant oversight. They lobby for EPA, the agency charged with overseeing the safe introduction of new pest-control components, to exercise stricter, more systematic control over GMO introductions. Saturating the environment with "Roundup Ready" or "Liberty-Link" crops, grown in rotation over multiple successive seasons, could stimulate herbicide-resistant weeds. Several successive seasons of these crops, carefully monitored, should provide some initial indications of safety or risk. In addition, it is already known that co-evolution of Bt-resistant pests is a clear and present danger. As a result, cultivation methods in Bt-insect resistant crops are carefully monitored, and farmers are officially advised by company and government advisories to leave a margin for non-resistant insects to

---

[26] Ibid.

[27] In 1998, University of Alabama at Auburn claimed to have made a breakthrough, by genetically engineering an herbicide-resistant tobacco plant through its chloroplast genome (Daniell, Datta, Varma, et al. 1998). In this method, foreign (resistance) genes are introduced into chloroplasts, not into the cell nucleus, and as a result, are not in and cannot be dispersed by pollen. Skeptics countered that the method would have very limited applicability because it remains difficult, almost impossible, to use the transformation method and successfully regenerate cells in cereals, a main crop category of interest.

reproduce. A problem with this procedure is that it is voluntary. Farmers must be willing to sacrifice a portion of their terrain (up to 50% in some Union of Concerned Scientists recommendations) to insect damage for this method of integrated pest management to work.

Both of these problems may prove more or less severe, depending on how many different strategies for pest control are in the ABT (seed) arsenal, and how many different varieties of seeds are available for farmers to "choose." Researchers point to new opportunities to pyramid Bt with other genetic sources of plant defense and enhance host-plant resistance through alternative strategies for escaping a pathogen's effect—e.g. modification of chemistry, morphology, or maturation schedule. But such GE solutions highlight the additional risk that seed selection, dominated by the ABT companies, will not, in the future, allow farmers to choose against herbicide-tolerant or insect-resistant varieties, the seeds of which deliver life science industry seed companies a premium because they control both seed development and distribution.

Early opponents of ABT argued that because of its impact on the structure of agricultural industry, ABT would reduce biodiversity by restricting choice. For the 1998 agricultural season, novel genes and traits have been integrated and approved in only a few varieties of a few species. But farmers still want them, and the resultant plunge in biodiversity would appear to be a major, albeit temporary, threat. Lack of choice—or misguided choice—could result in farmers sowing seed that may have an insect-resistant or herbicide-tolerant trait, but may not be the elite seed line best adapted to local growing conditions.[28]

Social and economic displacement and substitution effects

Early critics also anticipated major displacement and substitution effects in agriculture, manufacturing, food processing and trade because ABT, to an unprecedented extent, allows the nutrient contents of food plants to be manipulated for nutritional or industrial ends, or to express specialty chemicals. In the 1980s, high fructose syrups manufactured by new (ABT) enzyme processes caused significant displacements of cane sugar in world agricultural production, trade, and food processing, and raised the possibility that starchy cereals could be intersubstitutable in supplying raw material. Forecasters also predicted major market substitutions in fats and oils, especially as a result of new enzyme processing that converted lower-value vegetable oils into fats that mimicked cocoa butter.

---

[28] The discouraging results experienced by Mississippi Delta cotton farmers when they used Monsanto Roundup-Ready seeds supplied by Delta and Pine Land, may have been an artifact of this mismatch problem rather than of inclement weather and farmers' mistakes, which were Monsanto's explanation (Myerson 1998).

On a smaller scale, analysts predicted that laboratory-synthesized high-value chemicals, such as fragrances, flavors and insecticides, would displace natural raw botanical materials conventionally supplied by third world agriculture. Agriculturalists also noted that ABT held potential to introduce characteristics of wider environmental tolerance into important field and forage crops, and also to extend the ranges of tropical crops into temperate zones, and displace additional third world agricultural products.[29] Such predictions emphasize that substitution effects could be diverse and dynamic. They would vary by crop and location, and would cause new patterns of competition that could snowball from one effect to the next. Analysts also emphasize that technical change is one of three factors transforming food systems, and must be examined in the context of the relative costs of product substitutes. Change must also be examined in the context of political decisions that influence price and trade patterns. For example, US production of corn, and the relative price advantage of corn syrup over tariff-protected cane sugar, favors high substitution of the cheaper for the more expensive product. On the other hand, ABT feed products synthesized from crop wastes ("single cell protein"), though technically feasible, remain "uneconomic" and do not substitute for cheap feedgrains (such as sorghum) that undersell them.

We can see that a combination of technical, economic and sociocultural considerations underlie the limited ABT product substitutions observed thus far. But other techniques raise new possibilities. Plant cells excised from their normal location can be grown in flasks on artificial media, in a process called tissue culture. Such cells can be easily manipulated by changing the components of the medium or the environment in which they are grown. If such techniques were perfected for large-scale production, they would eliminate the need for the tropics to produce some crops. Indeed, socioeconomic assessments in the 1980s predicted that tissue culture products would replace traditional third world agricultural exports, such as chocolate, vanilla and chili. Early evidence suggests that, aside from sugar cane, which had already experienced a major plunge in demand in the 1980s, the largest displacement will be in cocoa, where high-laurate canola or soybean oils are replacing cocoa butter in the market. The cheaper vegetable oils mimic its properties and have the advantage that they can be grown by temperate zone farmers in the US and Europe. In the food industry, for example, recent debates have sprung up among British, Danish, French and Belgian chocolate manufacturers over the labeling of chocolate confections which use not 100% cocoa butter but instead substitute cheaper vegetable oils.

Technical predictions as to the future of such displacements may have been wrong, however. The major substitutions are projected to

---

[29] Junne, 1992.

come directly from plants genetically engineered to produce the higher-value, more desirable fatty-acid profile oils—not from enzyme processing of existing oils, as some technologists had forecast. Another prediction was that chocolate flavor produced in tissue culture would reduce demand for cocoa. In fact, cocoa cultivation in 1998 was reported to be in major decline, but crop analysts blame poor cultivation techniques, and increasing insect and disease loads in cocoa plantations, as contrasted with small-farmer biodiverse cocoa production strategies. Small cocoa farmers who find they cannot make a living from cocoa have sown other crops or left agriculture altogether. Price increases for cocoa might improve remuneration and small farmer investment in cocoa, but are unlikely, particularly if the major cocoa butter component loses out to cheaper fats.

Other potential substitution effects have not materialized for economic or related food safety reasons. Early ABT developments heralded better balanced and more nutrient-dense food crops. But high-lysine and high-methionine[30] maize and soybeans have not yet replaced conventional varieties. One promising strain of high-methionine, genetically-modified soybeans was discovered to be carrying an allergen, so had to be abandoned, and the experience may have retarded other developments. Fruits and vegetables still face hurdles comparable to those faced by conventionally-improved crops. For example, transformed squash and tomato varieties may still be attacked by diseases and insects not targeted by ABT protection; they may bruise easily, or have unrelated but unacceptable consumer characteristics. Too, people may be wary of eating GMOs, such as "new leaf" potatoes, out of a political or food-safety desire to avoid genetically-engineered products or the chemicals with which they are grown. Such factors have delayed acceptance of ABT and their corresponding displacement effects, though more displacements may occur as more products enter the marketplace.

Food safety

OECD evaluations of strategies for food safety identify severe gaps in knowledge of large-scale impacts, and emphasize a great need to improve systematic data collection, to move beyond a case-by-case method of evaluation. According to their 1994 expert report, safety evaluators need: (1) new toxicological concepts to assess novel foods, because the old animal feeding trials will not prove adequate; (2) better nutritional data on the composition of traditional foods;[31] and (3) new

---

[30] Lysine and methionine are essential amino acids required in a complete human diet.
[31] "In general, current data systems cannot supply detailed information on, for example, variation in the presence of natural toxins or specific macro-micronutrients in various tissues in different varieties of food species" (Kok, E.J. and H.A. Kuiper, 1996: Evaluation of Strategies for Food Safety Assessment of

immunotoxicity methods.[32] Feeding rats or other experimental animals "whole foods," the scientists discovered, may be counterproductive. Even rats need balanced diets; they suffer extreme digestive and other distress on diets high in the test product but absent of other nutrients.[33]

Moreover, the concept of "substantial equivalence" may prove quite useless in predicting which items need further testing. Disagreements persist on the possible dangers of accepting "no known risk" where ignorance of a similar product for comparison is the main problem. Certain plant families, such as legumes, differ widely in chemical products which would require further testing. And food plant databases to date contain only minimal information about levels of natural toxicants in particular species or varieties. The rise in fruit and vegetable consumption and vegetarianism, along with increasing consumption of "nutraceuticals" and "pharmafoods," raises the risk of consuming damaging quantities of toxins or allergens.

The protocol that established allergenicity of soybeans transformed with Brazil-nut protein lends a cautionary note. There were few susceptible individuals from which sera were available to test the allergenicity of the product. Moreover, as one group of researchers reported, "The majority of gene transfers in the field of plant biotechnology are from organisms that have no documented history of food use. These techniques will not be useful in predicting if these genes encode proteins that have the potential to be allergenic."[34] Other scientists disagree and argue that the risk of releasing novel allergens in genetically modified plants is remote because novel proteins can be evaluated by systematic comparison with known allergens.[35] More data and more discussion on these issues are needed.

In Europe, efforts are underway to create a common database.[36] Data might include both toxin and nutritional factors. One unintended boon of food safety regulation, which includes evaluation of the content of key nutrients, is better understanding of the variation in vitamin and mineral contents of different varieties of common fruits, vegetables, or

---

Genetically Modified Agricultural Products—Information Needs. In OECD Food Safety Evaluation, 1996). These weaknesses were demonstrated even in the few species-specific case studies presented at the 1994 workshop (OECD 1996).

[32] Kok and Kuiper, 1996.

[33] Hammond, Rogers, and Fuchs, 1996.

[34] Nordlee, Taylor, Townsend, et al., 1996.

[35] Evaluations can consider their molecular weight in the region of 10-70 kD, similarity of amino acid sequences to known allergens, and their resistance to digestion, to heat denaturation, glycosylation, and relative abundance. On this basis, "the assumption can be made that a protein is not likely to be allergenic unless its properties suggest the possibility." (Conclusion to OECD 1996.)

[36] Gry, Soborg, and Knudsen, 1996.

other foods. This would allow nutritionists to work with agricultural scientists to select more nutritious varieties of common crops. It would also allow dieticians and nutritionists to construct diets tailored to the nutritional needs and digestive capacities of select populations, such as the elderly.[37]

In summary, much research leading to regulatory action remains to be done. Viewed positively, ABT regulatory concerns present an opportunity to create and manage scientific nutritional knowledge in new and necessary ways. First, we need more systematic evaluation of nutritional constituents of diet and roles of particular foods in the diet; second, we need to study the kinds and levels of toxins, allergens, anti-nutritional factors and nutrients in particular plant food species, along with the possible effects of different modes of cultivation and processing. It remains unclear, however, through what mechanisms consumers and producers will control and assure this process.

Nature, society and farming: Social impacts on producers and consumers

The findings above suggest that farmers might have fewer seed choices as industry concentrates on particular products, and become more like factory-farm managers, controlling information and biochemical applications but allowed fewer and fewer choices of seeds and assorted agronomic packages. While large life-science industries—and some molecular biologists who seek to "save" farming—applaud this image of "factory" or "contract" farming, sociologists and environmentalists deplore it.[38] Factory farming is seen as destroying the traditional family farm, and also as creating artificial, enclosed environments susceptible to pathogens and other disturbances (especially as biodiversity declines).

The consumer also faces questions of control—directly over the products consumed and indirectly over the agricultural environment which produces the products. In 1998 US consumers let the authorities know that they were paying attention and concerned about the industrialization of their food supply. Like their European counterparts, the public responded overwhelmingly negatively to an initiative proposing to label as "organic" foods that were genetically engineered, sludge-nourished, or irradiated. The USDA, acknowledging this dissent, withdrew the proposal. USDA did not further address the controversial issue of labelling, however. As up to 20% of maize, 40% of soybeans, and 50% of cotton produced in the US come from genetically-modified seed, some of these products find their way into vegetarian and "health" foods, consumed by those who reject genetically engineered foods on

---

[37] Rosenberg, 1991.
[38] Goodman, Sorj and Wilkinson 1987.

principle. In 1998, US consumers cannot choose to avoid genetically modified foods unless they consume those labeled "organic"—these legally cannot contain genetically-engineered products.

In their vigorous response to the issues of "organic" food labelling, consumers and producers seem to be asking the government to provide clear guidance on organic and genetically-engineered qualities of foods through labelling. US regulators have several options. They might follow the lead of countries such as the UK, which in 1994 held a public consultation on ABT and in 1997 proposed the establishment of a single food standards board that could address food safety issues all along the food chain. US regulators might consider how to implement a proposal such as that in Norway, demanding that ABT product developers demonstrate how the product will contribute to sustainable development (how to define "sustainable development" is still under discussion). These are not so much anti-science and anti-technology proposals as ways to begin addressing the qualities of foods, how they are produced, and whether—through advisory boards or other mechanisms—the public might introduce factors other than corporate profitability into ABT priority-setting, now exclusively in the hands of industry.

Currently, consultation with the US public takes place only after the product has been developed and is close to commercial release. Similarly, World Bank publications on the "enabling environment" for biotechnology summarize model practices for planned ABT introduction mostly at the stage of field trials.[39] In their model case of Australia, the institutional biosafety committee of the product developer makes information available to a National Biosafety Committee, which then issues a safety advisory. The product developer publishes an announcement (press release), circulates information to "interested individuals and organizations registered ... for this purpose" and sends a description of the project to municipal authorities in the area where the release is to take place. The "public"—as they have designated these groups—then has thirty days to comment. All information presents the release as a scientific experiment, and no communications address larger questions of social context or effect. Although the exercise of "public consultation" is clearly aimed at creating an "enabling environment" that necessarily complements environmental biosafety, the procedures do not touch on "social impact" or try to assess social risks and benefits.

A real question is what new research and product development priorities might emerge if public oversight were introduced earlier into the regulatory process. It is unlikely that the US public would go in the direction of German-speaking countries, which tend to be most opposed to ABT for historical reasons. Amidst the rise of a well-supported "Green" party movement, they explain their opposition as a reaction

---

[39] Doyle and Persley, 1996.

44

against Nazi misuse of genetics during World War II (i.e. the eugenics movement). They also consider it a statement in favor of German romanticism, which envisions nature as something of high intrinsic value that must under no circumstances be tampered with.

Nor is the US public likely to move in the direction of France and Ireland, where the citizenry oppose ABT not only for ideological or religious reasons that scientists should not tamper with the natural order, but for more immediately practical reasons as well, particularly ABT's potential to destroy small farming. The low proportion (less than 2% in 1998) of the US population in farming professions means that the US public's connection with husbandry is less salient than in these European countries, where almost everyone interacts more closely with farmers. The anti-technology dimensions of the environmental ("Green") political movements are also less present in American experience.

In the US, by contrast, opposition to ABT appears to come from those who see ABT wrongly marginalizing the poor farmers, poor consumers, and those who would present alternative forms of change in the food system. Indeed, for Berkeley ecologist Miguel Altieri, who envisions ABT destroying biodiversity and rural social organization, the approach is much more an ecologist's struggle to preserve and present viable alternative farming methods. This reach toward alternatives is also recommended by philosophers and anthropologists who deplore the single-dimensionality of ABT, whose approach sees plant products and processes in narrow terms of genes, instead of whole plants and their meaningful place in people's ecosystems and diets.[40] These latter dimensions are shared even by those who diverge on just how sacred and untouched by genetic manipulation a plant should be, and see no intrinsic harm in passing a genetically engineered potato between their lips.

Conclusions

ABT presents multiple environmental, medical and genetic bioethical concerns and raises issues of environmental, social, nutritional and human-nature relational impacts. Positively, over the first decade, ABT developments demonstrate considerable scientific-technical achievements that are widely applicable in the nutritional and health sciences. They also have generated a repertoire of safety regulations increasingly shared among developed countries—seeking to demonstrate that ABT processes and products convey no harm. Negatively, measures and institutions to insure that ABT products convey positive social benefit have lagged, despite early emphasis on the need for greater accountability on the part of scientists, government and industry.[41]

---

[40] Richards and Ruivenkamp, 1996. Walgate 1990.
[41] See essays in Gaull and Goldberg, 1991.

The pace of ABT commercial product entry into the marketplace has been a few years slower than expected, but choices, as expected, have favored major field crops or species that are economically valuable and technically easy to manipulate. Concentration in what is now termed the "life-science" industries has, if anything, been more rapid and comprehensive than expected, perhaps fueled by the trend toward mergers and acquisitions in industries worldwide. So far, early choices have been dictated mostly by private sector firms, with an eye toward market opportunities tempered by regulatory hurdles. Following European experiences, especially in the UK,[42] US public advocates interested in influencing current and future trends in ABT could be shaping a more active role for public education and oversight, not only on risks and benefits, as is the case with outspoken biotechnology critics, such as Jeremy Rifkin, but also on priority-setting. So far, only Norway has tried to implement such a social code, by burdening product approval with demonstration of the impact on "sustainable development," a category which has yet to be defined clearly.[43] Additional European countries are innovating programs of rapid education and consultation. Two important questions for the future trajectories of ABT in the US and the rest of the world are: (1) what new agenda and research priorities—based on public rather than industrial interests—might develop from fresh discussions and advisory structures, and (2) how best to implement a process toward greater understanding.

References

Biotechnology Industry Organization 1998. Biotechnology Agricultural Products in the Market. Mailing. 25 March 1998 (see also www.bio.org).

Birch, R.G. 1997. Transformation: Problems and Strategies for Practical Application. *Ann. Rev. Plant Physiology and Plant Mol. Biol.* 48:297-326.

Cocking, E.C. 1990. All Sorts of Plant Genetic Manipulation. In Genetic Engineering of Crop Plants. G.W. Lycett and D. Grierson, Eds. pp. 1-12. London: Butterworth.

Daniell, H., R. Datta, S. Varma, S. Gray, and S-B. Less 1998. Containment of herbicide resistance through genetic engineering of the chloroplast genome. *Nature Biotechnology* 16 (April): 345.

Doyle, J.J. and G.J. Persley 1996. Enabling the Safe Use of Biotechnology: Principles and Practice. World Bank Environmentally Sustainable Development Studies and Monographs Series No. 10. Washington, DC.

---

[42] UK 1994.
[43] Sandberg 1995.

46

FAO/WHO 1997 Codex Alimentarius.

Fuchs, R.L., D.B. Re, S.G. Rogers, B.G. Hammond, and S.R. Padgette 1996. Safety evaluation of glyphosate-tolerant soybeans. In OECD, Food Safety Evaluation, pp. 61-70.

Gaull, G.E. and R.A. Goldberg 1991. New Technologies and the Future of Food and Nutrition: Proceedings of the First Ceres Conference. Williamsburg, VA, October 1989. New York: John Wiley and Sons, Inc.

Goodman, D., B. Sorj, & J. Wilkinson 1987. From Farming to Biotechnology: A Theory of Agroindustrial Development. New York: Blackwell.

Gry, J., I. Soborg, and I. Knudsen 1996. The role of databases: The example of a food plant database. In OECD, Food Safety Evaluation, pp. 118-29.

Hammond, B., S.G. Rogers, R.L. Fuchs 1996. Limitations of whole food feeding studies in food safety assessment. In OECD, Food Safety Evaluation, pp. 85-97.

Hobbelink, H. 1991. Biotechnology and the Future of World Agriculture: The Fourth Resource. London: Zed Books.

Hodgson, J. 1992. Biotechnology: Feeding the World? Bio/Technology 10.

James, C. and Krattinger 1996. Global Review of the Field Testing and Commercialization of Transgenic Plants, 1986 to 1995: The First Decade of Crop Biotechnology. Ithaca, NY: ISAAA.

Junne, G. 1992. The Impact of Biotechnology on International Commodity Trade. In Biotechnology: Economic and Social Aspects: Issues for Developing Countries. E.J. DaSilva, et al., Eds. Cambridge: Cambridge University Press.

Kilman, Scott and Susan Warren 1998, 27 May. Old Rivals Fight for New Turf—Biotech Crops. Wall Street Journal B1.

Kok, E.J. and H.A. Kuiper 1996. Evaluation of strategies for food safety assessment of genetically modified agricultural products—Information needs. In OECD, Food Safety Evaluation, pp. 80-84.

Krimsky, S. and R.P. Wrubel 1996. Agricultural Biotechnology and the Environment: Science, Policy and Social Issues. Urbana, IL: University of Illinois Press.

Lazarus 1996. In OECD, Food Safety Evaluation, pp. 98-106.

Messer, E. 1992. Sources of Institutional Funding for Agrobiotechnology for Developing Countries. ATAS Bulletin (United Nations Center on Science, Technology and Development) 9 (Winter 1992): 371-79.

Messer, E. & P. Heywood 1990. Trying technology: Neither sure nor soon. Food Policy (August 1990): 336-345.

Mikkelson, T.R., B. Anderson, and R.B. Jorgensen 1996. The risk of crop transgene spread. Nature 380:31

Monsanto 1998. Monsanto Company 1997 Annual Report. St. Louis, MO.

Myerson, A.R. 1998. Monsanto settling genetic seed complaints. Feb. 24 *New York Times*, p. D2.

Nordlee, J.A., S.L. Taylor, L.A. Thomas, and J.A. Townsend 1996. Investigations of allergenicity of Brazil nut 2S seed storage protein in transgenic soybean. In OECD, Food Safety Evaluation, pp. 151-155.

OECD 1992. Safety Considerations for Biotechnology. Paris.

OECD 1993. Safety Evaluation of Foods Derived through Modern Biotechnology: Concepts and Principles.

OECD 1994. Aquatic Biotechnology and Food Safety.

OECD 1996. Food Safety Evaluation.

Richards, P. and R. Ruivenkamp 1996. New Tools for Conviviality: Society and Biotechnology. In Nature and Society: Anthropological Perspectives, P. Descola and G. Palsson, eds. NY: Routledge Press.

Rissler, J. and M. Mellon 1993. Perils Amidst the Promise: Ecological Risks of Transgenic Plants in a Global Market. Cambridge, MA: Union of Concerned Scientists.

Rosenberg, I. 1991. Nutrition and the Elderly. In New Technologies and the Future of Food and Nutrition. G.F. Gaull and R.A. Goldberg, Eds., pp. 79-82. New York: John Wiley.

Sandberg, Per, Ed. 1995. Release and Use of Genetically Modified Organisms: Sustainable Development and Legal Control. Proceedings of the International Conference organized by the Norwegian Biotechnology Advisory Board 1995. Oslo, Norway, 13-14 September 1995.

Sasson, A. 1988. Biotechnologies and Development. Paris: UNESCO. Technical Center for Agricultural and Rural Cooperation (CTA).

Thro, A.M., N. Taylor, C. Raemakers, et al. 1998. Maintaining the Cassava Biotechnology Network. Nature Biotechnology 16 (May): 425-27.

UNCED (United Nations Conference on Environment and Development) 1992 Agenda 21: Convention on Biodiversity.

UK (United Kingdom) National Consensus Conference on Plant Biotechnology 1994 Final Report. London: Science Museum.

Walgate, R. 1990. Miracle or Menace? Biotechnology and the Third World. London: The Panos   Institute

Table 1, part I.

Most Common Methods of Genetic Transformation of Plant Cells

(Less commonly used methods of transformation include polyethylene glycol (PEG)-mediated uptake of DNA by protoplasts, and microinjection of DNA into cells. The PEG method seems to be non-species-specific, but requires the use of protoplasts. Microinjection can be used to transform only one cell at a time, and is labor intensive.)

Agrobacterium tumefaciens-mediated transfer	Electroporation of protoplasts	Microprojectile bombardment
Method	**Method**	**Method**
• First described in 1983.	• First reported in 1989.	• First reported in 1988.
• Tumor-inducing (Ti) plasmid of bacterium is disabled by removing genes responsible for onset of crown gall disease.	• Electrically-mediated uptake of DNA by specially prepared plant cells (protoplasts)	• DNA is coated onto μm-sized metal particles suspended on a macroprojectile.
• Remaining Vir genes facilitate excision of T-DNA region from plasmid and integration of region into host DNA.	• Protoplasts are prepared by removing cell walls of plant cells.	• The macroprojectile is discharged towards a stopping plate positioned above plant tissue.
• Genes of interest are inserted into T-DNA region.	• These are subjected to rapid electrical stimulus, temporarily rendering membrane permeable to DNA.	• Microprojectiles continue directly into plant tissue.
• Binary vector system may be used, in which genes flanked by T-DNA repeats are located on a second, non-Ti plasmid.		

Table 1, part 2.

Species and characteristics

- Natural targets are dicotyledonous plants, including tobacco, tomato, cucurbits.
- Monocotyledons have proven more difficult to transfect by this method. This method may be preferred when transfected plants will be regenerated through embryogenesis rather than meristem or tissue culture.
- Has been used to introduce viral resistance coat-protein genes.

Regulatory issues

- Ti vector sequences derived from a plant pathogen must be viewed as potentially pathogenis. For deregulated status, developers must prove that presence of T-DNA does not cause tumorigenicity.
- Results in precise, limited-copy integration of transgenes in genome.

Species and characteristics

- Cereals have often proved difficult to regenerate from protoplasts.
- Has been used to introduce disease and insect resistance phenotypes.

Regulatory issues

- No infectious agents of viral DNA are used in this method.

Species and characteristics

- Has been shown to be applicable to virtually all species, with varying ease. Is the only method for delivering DNA into mitochondria and chloroplasts.
- Has been used to introduce insect resistance, herbicide tolerance, and product quality phenotypes.

Regulatory issues

- May lead to nuclear rearrangements of the introduced DNA.
- May result in fragmented copies of genes or vectors at different sites in genome.

GENETIC PREDISPOSITION
AND THE POLITICS OF PREDICTION
Dorothy Nelkin

Throughout history, shamans and soothsayers, astrologers and oracles, wizards and witches have read, assessed, and thereby tried to tame the future. In the millennial fervor of the end of the 20th century, we are virtually flooded with predictions—from futurists, publicists, scientists, and especially these days, geneticists, as their discoveries are enhancing the ability to predict future health and disease. The tests emerging from the science of genetics can detect inner biological conditions that are predictive of possible future diseases in people who are currently healthy, expressing no symptoms.[1] Such tests can identify predisposition not only to Mendelian disorders such as Huntingdon's Disease or Cystic Fibrosis, but also to a growing number of common and complex conditions including various forms of breast and colon cancer or Alzheimer's disease. Moreover, the concept of genetic predisposition is often extended—especially in popular culture—to the explanation and anticipation of a growing range of behavioral characteristics and personality traits.

Geneticists have encouraged futuristic fantasies about the predictive power of their field. In their public statements and popularizations, they have called the human genome a "blueprint for destiny," a "Delphic Oracle," a "time machine," a "trip into the future," a "medical crystal ball." Nobelist and first director of the U.S. Human Genome Project James Watson has claimed in public interviews that "our fate is in our genes."[2] Capturing the mood of this science, a cartoon portrays Madame Rosa, an astrologer, standing in her shop next door to Madame Tina, a geneticist. Both are in the business of predicting future fate.

Prediction is believed to be the essence of science; the common definition of a successful theory is its ability to predict. But numerous historical studies have documented the powerful influence of social context and political ideology on many scientific ideas, suggesting that some ideas are accepted not only for their explanatory power, but for the legitimacy they provide for political and social agendas.[3] The politics of prediction raises troubling and controversial questions when applied to

---

[1] For details on genetic testing, see Neil Holtzman, <u>Proceed with Caution</u>. Baltimore: Johns Hopkins Press, 1989. Also see the National Institutes of Health, Task Force on Genetic Testing, <u>Report</u>, Bethesda, MD, 1997.
[2] Watson is quoted in Leon Jaroff, "The Gene Hunt," *Time*, March 20, 1989. 62-67.
[3] See, for example, Robert Proctor, <u>Value Free Science</u>. Cambridge: Harvard University Press, 1991.

science—presumably a neutral and objective endeavor and, therefore, a value-free guide to policy decisions. Are science-based predictions valued because they are likely to be true? Or are there other reasons for the appeal of certain predictions—for example, their utility as a confirmation of prevailing stereotypes or as a legitimization of particular social policies? How are scientific predictions appropriated in public affairs? This paper addresses such questions, focusing on the widespread popular and policy appeal of theories purporting to predict behavioral predispositions.

First, it is useful to clarify the concept of predisposition—a term which has both lay and scientific meanings. The meaning of predisposition in science or medicine is often quite different from its usage in social context. In the clinical setting *a predisposition is not a prediction, but a statistical risk calculation.* Identifying a predisposition assumes the existence of a biological condition signaling, on the basis of probability, that an individual may eventually suffer a disease or behavioral aberration. A person predisposed to cancer, for example, may have a biological quality (the BRCA-1 gene) that heightens the odds of getting cancer, just as someone who drives every day has heightened odds of being involved in an accident. But as we know from the debates over the actual meaning of BRCA-1 and its implications for the individual at risk, many variables influence the occurrence of the actual event. Thus, in its clinical meaning, to be predisposed is to be vulnerable to a disease or a condition that may or may not be expressed some time in the future.

In its social meaning, however, this statistically driven concept of predisposition often becomes a definitive prediction. And possible future states, calculated by statistical methods, become equivalent to current status. In media reports, for instance, a genetic predisposition appears as immutable, a powerful force in shaping physical status and social behavior, and a predictable part of the human condition. A genetic aberration that *may contribute to a predisposition* to obesity or alcoholism becomes an "obesity gene" or an "alcohol gene." The BRCA-1 mutation has become the "breast cancer gene." And the predisposed individual becomes labeled as a "risk." Although most scientific research has focused on predisposition to diseases with a known genetic component, the popular interest in genetic predisposition has focused less on disease than on behavior—especially problematic behavior such as violence or alcoholism—as people relate genetics to their most urgent social and political concerns.

In The DNA Mystique: The Gene as a Cultural Icon, historian M. Susan Lindee and I document the popular appeal of genetic explanations in mass culture.[4] Exploring diverse media, including popular magazines, television sitcoms and soaps, newspaper reports, advertisements, child care books, and films, we found hundreds of articles and stories about genetic predisposition to diseases and to an astounding range of behaviors and personality traits. Among the traits attributed to genes in the popular media, some have been more plausible than others: they include homosexuality, criminal violence, aggressive tendencies, addiction, exhibitionism, arson, intelligence, learning disabilities, tendency to tease, propensity for risk taking, family loyalty, religiosity, social potency, tendency to giggle, traditionalism, happiness, and zest for life. Media stories of genetic predispositions are mainly preoccupied with deviant behavior, and with the possibility of predicting and thereby controlling anti-social tendencies by identifying the predisposed.

Media matters. Repeated messages in mass culture both reflect and create the widespread and often unarticulated assumptions that underlie social policy. I therefore focus on widely circulated media narratives that convey assumptions about the biological causes of deviance, and I will suggest the similarities between these contemporary stories and those that circulated during the eugenics movement some 80 years ago.

Predicting deviant behavior

In 1912, a eugenicist, G. Frank Lydston, wrote a play called The Blood of the Fathers.[5] It was about a young doctor, interested in hereditarian problems, who falls in love with a charming patient, supposedly the child of an Army officer. Just before their marriage, the doctor discovers that Kathryn, his bride-to-be, is actually the biological daughter of a convicted murderer and an opium-addicted mother who had committed suicide. Faced with the "horrible outlook which only his knowledge of degeneracy and hereditary criminality can give," he experiences "an awful conflict of emotions." But "the man triumphs over the scientist." He ignores his science-based predictions and marries the woman he loves.

Soon after their marriage, however, his bride notices a valuable ornament in the hair of an elderly woman and "instinctively" filches the jewel. "Her blood tells," writes Lydston. Confronted with her crime, Kathryn "goes to pieces" and kills herself, leaving the doctor convinced

---

[4] Dorothy Nelkin and M. Susan Lindee, The DNA Mystique: The Gene as a Cultural Icon. New York: Freeman, 1995.
[5] G. Frank Lydston, The Blood of the Fathers: A Play in Four Acts Dealing with the Heredity and Crime Problem. Chicago, Riverton Press, 1912. Lydston also published a popular book on criminality, The Diseases of Society (Philadelphia: Lippincott, 1912).

that "love never can survive mismating." In the final scene, the doctor summarizes the moral of the story: "Poor Kathryn! You were wiser than you knew. You set things right—and you did it the only way... The blood of the fathers! And our children yet unborn—and our children's children, they too, thank God, are saved!... and in the only way."

Today, the "blood of the fathers" has been replaced by the "genes," as the predisposing agent, and the old eugenic concept of biological determinism has been replaced by the label, "genetic predisposition." But the stories have hardly changed at all. Take, for example, the plot of a 1993 prime time television play called Tainted Blood.[6] A 17-year-old boy from a "stable" family unexpectedly kills his parents and then himself. The case attracts an investigative reporter (Raquel Welch) who discovers that the boy and his twin sister had been adopted at birth because their genetic mother was in a mental institution. When the reporter finds out that the mother, like the son, had killed her parents and then herself, she begins a frantic search for the twin sister and her adoptive family. For, it is assumed, the twin sister would also be predisposed to kill. Predictably, after threatening her adoptive family, the twin does kill herself. Like Lydston's play over 80 years ago, this contemporary story concludes with a moral message: the girl was not to be blamed for her actions, for she had inherited "a genetic disease": she was "predisposed."

The idea of genetic predisposition has had a persistent history. In the 1960s the media published many stories on the so-called "criminal chromosome" as an explanation and predictor of deviance. The British cytogeneticist Patricia Jacobs had found that a disproportionate number of men in an Edinburgh correctional institution were XYY males, and she suggested that the extra Y chromosome "predisposes its carriers to unusually aggressive behavior."[7] Further research challenged Jacobs' initial study, but the "criminal chromosome" enjoyed a long an remarkably popular life. It attracted the attention of the press in April 1968 when it was invoked to explain one of the most gruesome crimes of the decade. A *New York Times* reporter wrote that Richard Speck, then awaiting sentencing in the deaths of 9 student nurses, planned to appeal his case on the grounds that he was XYY. This story, which was, as things turned out, incorrect (Speck was an XY male) provoked a public debate about the biological causes of criminal behavior. *Newsweek* asked

---

[6] Tainted Blood. USA Channel, March 3, 1993, 9:00-11:00 pm.
[7] P.A. Jacobs, M. Bruton, M.M. Melville, R.P. Brittan and W.F. McClement, "Aggressive behaviour, mental subnormality and the XYY male." *Nature* 208 (1965) 1351-2.

if criminals were "Born bad" – "Can a man be born a criminal?" and *Time* headlined a story "Chromosomes and crime."[8]

References to the criminal chromosome have continued to enter popular views of criminal violence. News reporters and talk show hosts refer to "bad seeds," or "criminal genes."[9] To another *New York Times* writer, evil is "embedded in the coils of chromosomes that our parents pass to us at conception."[10] Media stories like Tainted Blood or various versions or Born To Kill suggest that those individuals who are "born to kill" will predictably do so, despite their comfortable home life or favorable social advantages.

In December 1991, a fourteen-year-old high school boy was arrested for the murder of a schoolmate. The *New York Times* account of this event interpreted it as a key piece of evidence in "the debate over whether children misbehave because they had bad childhoods or because they are just bad seeds...." The boy's parents had provided a good home environment, the reporter asserted; they had "taken the children to church almost every Sunday." Yet their son had been arrested for murder. This troubling inconsistency between the child's apparently decent background and his violent behavior called for an explanation. The reporter resolved the mystery through the power of predisposition: the moral of the story was clearly stated in the article's headline: "Raising Children Right Isn't always Enough." The implications? There are, indeed, "bad seeds."[11]

The interest in "bad genes" also appears in speculations about the nature and etiology of addiction, and especially in stories about "alcohol genes."[12] Definitions of alcoholism have shifted over time from sin to sickness, from moral transgression to medical disease, depending less on changes in scientific understanding than on prevailing social, political and moral agendas.[13] Early leaders of the American temperance movement defined alcoholism as a disease, but when the movement began to advocate outright prohibition, alcoholism was redefined, along

---

[8] Cited in Jeremy Green, "Media Sensationalism and Science: The Case of the Criminal Chromosome" in Terry Shinn and Richard Whitley, Eds., Expository Science, Sociology of the Sciences Yearbooks, Vol. IX, 1985, 139-161.

[9] See, for example, the language used by James Fallows in "Born To Rob? Why Criminals Do It," *The Washington Monthly*, December 1985. 37.

[10] Deborah Franklin, "What a Child is Given," *New York Times Magazine*, September 3, 1989. 36.

[11] Maria Newman, "Raising Children Right Isn't always Enough," *New York Times*, December 22, 1991.

[12] George Nobbe, "Alcoholic Genes," *Omni*, May 1989. 37.

[13] Sheila B. Blume, M.D., "The Disease Concept of Alcoholism," 1983 *Journal of Psychiatric Treatment and Evaluation*, Vol. 5, 417-478. David Musto, The American Disease. New Haven: Yale University Press, 1973; Joseph Gusfield, The Culture of Public Problems. Chicago: University of Chicago Press, 1981.

with syphilis and opiate addiction, as a "vice"—a manifestation of immoral behavior. Then, Alcoholics Anonymous, founded in 1935, insisted that alcoholism was a compelling biological drive that could be cured only by total abstinence and moral rectitude. This organization, seeking to remove the stigma from the condition, strongly promoted the idea that alcoholics had "predisposing characteristics" that distinguished them from others.[14] This view has been reinforced by genetic explanations of addiction.

Common observation shows, of course, that alcoholism does run in families. As in the case of violence, however, this in itself does not necessarily imply that it is a hereditary condition. The prevalence of alcoholism in certain families could reflect role models, or an easy availability of alcohol, or a reaction to abuse. Nevertheless, popular magazine articles have expressed a common perception: "Addicted to the bottle? It may be in your genes."[15] Or as *Mademoiselle* asked: "Do you have a gene that makes you a designated drinker?"[16] Writing about the "natural bent" of alcoholics, *New York Times* reporter Dan Goleman suggested one reason for the appeal of genetic explanations: Identifying an "alcoholism gene" offers the hope that addiction can be predicted and, therefore, controlled, not through the uncertain route of social reform, but through biological manipulation.

Assumptions about genetic predisposition to alcoholism have extended to other addictions: smoking, overeating, shopping, and gambling. A *New York Times* story claims: "Smoking has to do with genetics, and the degree to which we are all prisoners of our genes.... You're destined to be trapped by certain aspects of your personality. The best you can do is put a leash on them."[17] Books about obsessive compulsive behavior describe the "victims of a disease process;" "Most researchers agree OCB will develop only if an individual is genetically predisposed to it."[18]

To explain such behaviors in the absolute terms of genetic predisposition is to extract them from the social setting which defines and interprets behavior. There are, of course, no criminal genes or alcohol genes, but only genes for the proteins that influence hormonal and physiological processes. And only the most general outline of social

---

[14] Peter Conrad and Joseph Schneider , Eds., Deviance and Medicalization. St Louis: C.V. Mosby, 1980.
[15] George Nobbe, "Alcoholic Genes," *Omni*, May 1989. 37.
[16] Shifra Diamond, "Drinking Habits May Be in the Family," *Mademoiselle*, August 1990. 136.
[17] Laura Masnernus, "Smoking: Is it a habit or is it genetic?" *New York Times Magazine*, October 4, 1992.
[18] Lee Bar, Getting Control: Overcoming Your Obsessions and Compulsions. New York: Plume, 1992. 12-13.

behavior can be genetically coded.[19] Even behaviors known to be genetically inscribed, such as the human ability to learn spoken language, *do not appear if the environment does not promote them.* In the case of addiction, any biological or genetic predisposition that may exist would only become a full-blown pattern of behavior in an environment in which alcohol was readily available and socially approved.

However, many interests are at stake in perpetuating a belief in individual predisposition as the cause of deviant behavior, for causal explanations imply moral judgments about responsibility and blame. If defined as a sin, alcoholism represents an individual's flouting of social norms; if defined as a social problem, it represents a failure of the social environment; if defined as intrinsic to the product, it represents the need for regulation. But if defined as a genetic predisposition, *neither the society, nor the industry appear responsible.* And even the addicted individual cannot really be blamed. The practical importance of this is becoming increasingly evident in several policy contexts.

In the political climate of the 1900s—in particular the efforts to dismantle the welfare state—the idea of genetic predisposition and the translation of predisposition into prediction hold significant policy appeal. Individual predispositions can be appropriated, for instance, to explain social inequities without blaming public policies or social institutions, and predictions about their influence can be used to support the popular policy position of the 1990s—that there are limits to social intervention.

Explaining social inequities

During the eugenics movement in the late nineteenth and early twentieth century, conditions such as "pauperism" were defined as "in the blood." This was the conclusion of common observations and they were supported by the science of the time. For, after all, poverty persists in families over several generations.[20] Scientists did field research to identify "pathological" families that were predisposed to specific disabilities such as laziness, harlotry, loquacity, or lack of impulse control. In retrospect, it is easy to see the fallacies in such formulations and generalizations. Yet similar beliefs have re-emerged in recent years in public discourse, appearing in discussions and debates about what makes people different. The idea of genetic predisposition has become an expedient way to explain the paradox of poverty in an affluent society. People are simply driven—and limited—by their genes.

In a culture obsessed with fame, money, and personal power, the cause of exceptional achievement is also a matter of curiosity. Why are

---

[19] Richard C. Lewontin, Biology as Ideology. New York: Harper and Row, 1992. 51.
[20] Daniel Kevles, In the Name of Eugenics. New York: Knopf, 1985.

some people more successful than others? What accounts for extraordinary achievements? There are many possible explanations for success: hard work, persistence, talent, exposure to role models, availability of professional opportunities, social pressures from family or peers, or simply good luck. But another set of narratives suggest that achievements have genetic origins, that some are predisposed to be successful while others fail. Just as children from good families may turn out bad, so those with limited opportunity—if they have the proper genes—will, predictably, rise above their social circumstances. Thus, a television newscaster, describing a teenager from a deprived background who became captain of his track team and won a college scholarship, remarked on the source of the boy's achievements: "He has a quality of strength and I guess it has a genetic basis."[21] A *Newsweek* article explained more generally how poverty, physical impairment or abuse affects children differently: "Some kids have protective factors that serve as buffers against the risks." They have "natural resilience" or "built-in defenses." It is the "genetic luck of the draw."[22]

Stories about popular personalities, be they scientists, actors, sports heroes, politicians, or rock musicians, often refer to their genetic predispositions. We read about Elvis's genes and Einstein's DNA. The range of special talents attributed to genetic predisposition is remarkable. In a story about the Ginsberg brothers, both of whom are poets, the *New York Times* referred to their "poetry genes."[23] An obituary writer explained the secret of Isaac Asimov's success: "It's all in the genes."[24] The idea of genetic predisposition appears in unlikely places. A horoscope columnist announced that he was "genetically coded with impeccable social instincts."[25] A gardener celebrated her "gardening genes."[26]

Genetic explanations are especially convenient in the context of current educational and welfare policies. In the 1990s, neo-conservative critics of liberalism published books and essays explaining social distinctions among various groups in American society in terms of

---

[21] NBC News Special, "Kids and Stress," April 25, 1998.

[22] David Gelman, "The Miracle of Resiliency," *Newsweek* Special Issue. Summer 1991. 44-47.

[23] Barbara Delatiner, "For Brothers, Poetry is in Their Genes," *New York Times*, May 26, 191.

[24] Mervyn Rothstein, "Isaac Asimov, whose Thoughts and Books Traveled the Universe, Is Dead at 72," *New York Times*, April 7, 1992. B7.

[25] Rob Brezsny, "Real Astrology," *New York Press*, August 5-11, 1992. 54.

[26] Martha Smith, "I Found it in my Other Genes," in Beds I Have Known: Confessions of a Passionate Amateur Gardener. New York: Atheneum, 1990. 17.

intrinsic biological differences.[27] Their policy agenda was to eliminate liberal social welfare programs. They claimed that scientific evidence demonstrated the importance of genetic differences, and suggested there were biological limits to social intervention. The most widely discussed of these critiques was the best-selling book, The Bell Curve, by Richard Herrnstein and Charles Murray.[28] They made claims about the critical importance of genetically determined differences in intelligence among races in perpetuating and justifying social class differences, and they called for social policies that recognized biological distinctions and predispositions.

Such arguments have been widely discussed and often cited in policy debates. They have appeared, for example, in immigration debates as an argument for restricting the immigration of specific groups. In a popular anti-immigration book, Peter Brimelow posits a Darwinian view of world history in which the development of certain national traits, he claims, are "to a considerable extent biological."[29] Biological arguments also appear in a statement proposing new guidelines for future philanthropy. Private philanthropy has long been based on the conviction that, given the opportunities provided by money, people can change. But according to this new statement, written as a response to The Bell Curve, there is evidence ("widely accepted by experts") about the heritability of intelligence and other behavioral characteristics that challenges the "old" conviction. "Philanthropic efforts to help disadvantaged groups may well be thwarted to the extent that their differences are hereditary."[30]

This appropriation of DNA to explain individual differences recasts old and pervasive beliefs about the importance of "blood" in powerful and contemporary scientific terms. Science has long served as a way to support the status quo as "natural," that is, dictated by natural and inexorable forces. The concept of predisposition today is also used to uphold existing social categories, whether based on gender, race, or economic circumstances, as inevitable. Race or gender inequalities can be explained in terms of biological fate. The great, the famous, the rich and successful, are what they are because of their genes. So too, the deviant, the dysfunctional and the poor are genetically fated. Opportunity

---

[27] See, for example, Daniel Seligman, A Question of Intelligence. New York: Carol Publishing Corp., 1992; and J. Phillippe Rushton, Race, Evolution and Behavior. New Brunswick: Transaction Books, 1994.

[28] Richard Herrnstein and Charles Murray, The Bell Curve. New York: The Free Press, 1994.

[29] See Peter Brimelow, Alien Nation. New York: Random House, 1995. For a review of the use of genetic arguments in contemporary immigration debates, see Dorothy Nelkin and Mark Michaels, "Biological Categories and Social Controls," forthcoming 1998.

[30] Lemkowitz, Leslie, "What Philanthropy can learn from the Bell Curve," Hudson Institute, November 29, 1994.

is less important than predisposition. Some are destined for success, others for problems, or at least a lesser fate. A star—or a criminal—is not made but born.

This is a particularly striking theme in American society, where the very foundation of the democratic experiment was premised on the belief in the improvability—indeed, the perfectibility—of all human beings. The very idea of immutable predisposition contradicts the "bootstrap ideology" that has pervaded American folklore, undermining the myth that an individual's will or hard work alone determines his or her achievement or success. The belief in genetic destiny implies natural limits constraining the possibilities for both individuals and social groups. Humankind is not perfectible because the species' flaws and failings are inscribed in an unchangeable text—the DNA—that will persist in creating murderers, addicts, the insane and the incompetent, even under the most ideal social circumstances. In popular stories, children raised in ideal homes become murderers, and children raised in difficult home situations become well-adjusted high achievers. In child care books today parents are counseled on the importance of recognizing the predispositions of their children. The moral of these narratives? Neither individual actions nor social opportunity really matter if our fate lies in our genes. There is no possible ideal social system, no possible ideal nurturing plan that can prevent the deviant acts that seem to threaten the social fabric of contemporary American life. For "predispositions" are a problem of the individual, absolving the political system and the broader culture from responsibility and blame. Such beliefs have opened the way to the policy appropriation of genetic predictions.

Responsibility and blame

I have suggested above that contemporary discussions of "criminal genes"—of biological predisposition to violence—sound much like the early hereditarian beliefs about "criminal types" that had so appealed to G. Frank Lydston in his play about choosing a mate. Such beliefs during the eugenic movement had also motivated the study of "pathological families" suspected of predisposition to a remarkable range of "deviant" traits. Then and now, hereditarian explanations of deviant behavior—in both professional and popular discourse—are important in defining responsibility and locating blame for social problems.

These explanations now appear in various institutional contexts. In the courts, for example, the concept of genetic disposition has been translated into "the genetic defense" and used to define the limits of criminal responsibility and free will. Appearing to be scientifically grounded, and more specific than the insanity defense, the genetic defense is appealing in the courtroom.[31] In 1995, geneticist Xandra Breakfield associated the high incidence of impulsive aggression among men in a Dutch family with an extremely rare mutation in the gene for the MAO enzyme. Though this was considered a highly unusual situation, the publication of her data stimulated lawyers to develop genetic defenses in cases of violent crime. For example, the lawyers appealing the Georgia death sentence of a murderer, Stephen Mobley, used a genetic defense to argue that the defendant was not responsible for the crime because his genes predisposed him to violence. Mobley did come from a family line which included a number of aggressive males (though some had expressed aggression in successful business ventures). His lawyers suggested this could imply a predisposition to violence. Mobley, they argued, was not responsible; he could not control his actions. They asked the court to give their client a genetic test, but the court refused to pay the costs and the lawyers eventually dropped the argument.

Mobley's lawyers claimed their client was genetically predisposed to violence in order to mitigate his punishment. But biology-based arguments are malleable, and could easily be appropriated for other ends. The perception that genetic conditions are hopeless and immutable could call, for example, for permanent incarceration or even the death penalty for those with "bad genes."

The tendency to blame antisocial behavior on biological predisposition reflects the apparent intransigence of social problems and the public disillusionment about the efficacy of social policies founded on environmental determinism. And it suggests the appeal of science-based predictions as a way to predict and prevent violent crime. "Rape could be reduced greatly," says a writer for the American Airlines magazine, *The American Way*, "if we had a way to determine who was biologically predisposed to it and took preemptive action."[32] Some scientists have also suggested the policy implications of predictive information about criminal tendencies. A molecular biologist and former editor of *Science* describes a wide variety of violent crimes and then suggests: "When we can accurately predict future behavior, we may be able to prevent the damage."[33] Recently some states have enacted sexual predatory statutes that require "propensity hearings." In the context of fear of crime, scientifically based predictions hold considerable—and dangerous—appeal.

---

[31] Rochelle Dreyfuss and Dorothy Nelkin, "The Jurisprudence of Genetics," *Vanderbilt Law Review* 45, 2, March 1992, 313-348.

[32] Joel Keehn, "The Long Arm of the Gene," *The American Way*, March 1992, 36-38, citing William Thompson from the Univ. of California.

[33] Daniel Koshland, "Elephants, Monstrosities and the Law," *Science*, 255, February 4, 1992. 777.

Given the pressures of cost and time that currently plague the criminal justice system, genetic explanations of violent behavior also fit conveniently with current ideologies about prison reform. Disillusioned with the failure of past rehabilitation schemes and pressed to save money, criminologists are learning toward the "selective incapacitation" of prisoners instead of efforts to rehabilitate them.[34] Theories of behavioral genetics can be appropriated to justify these trends. One psychologist even says, "The criminal is a different species entirely... there is nothing to which to rehabilitate a criminal."[35] Research funds in the field of criminology are increasingly directed toward studies of the biological cause of violent behavior.

Claims about predisposition have the rhetorical advantage of being both unfalsifiable and irrefutable. To say, for example, that a man committed a crime because he was predisposed to criminal behavior can never be disproven, for the outcome is proof of its own cause. Thus, a claim of genetic predisposition can be invoked to explain any behavior; an individual will respond either poorly or well to a particular environment depending on pre-existing biological tendencies. Some have "built-in resistance;" others have "evil in the genes." In fact, whatever the influence of genetics, biological predisposition is not necessary to explain why an inner-city African American, growing up in a climate of racism, drug abuse, and violence, and without much hope of escaping that climate, might become indifferent to human life. Yet it is expedient—especially to those concerned about the public costs of social programs—to believe that problems rest less with society than with individual predispositions.

Claims about genetic predisposition are taking place in other policy arenas where they are also used as a way to shift responsibility and locate blame. Cancer is increasingly defined as a genetic disease, an inherited predisposition. Though cancer is, indeed, a genetic disease in the sense that it involves gene mutations, not all types of cancer are inherited. Environmental influences are responsible for many mutations. But the redefinition of cancer as a genetic disease shifts blame and responsibility away from industry and regulators.

It is also convenient to attribute addiction, not to products but to the individuals who are predisposed. Gallo Wine, for instance, is supporting research on alcoholism at the Ernest Gallo Clinic and Research Center. Gallo scientists have located a gene that produces a protein that, they claim, jams the signals warning a person to stop drinking. Critics, however, note that genetic explanations are useful to

---

[34] C. Ray Jeffery, Criminology: An Interdisciplinary Approach. New Jersey: Prentice Hall, 1990 184.

[35] Jakes Page, "Exposing the Criminal Mind," Science 84, September 1984. 84-85.

the industry, locating responsibility for alcoholism in certain individuals. This would mean that others need not worry about how much they drink.[36] Similarly, the tobacco industry has supported research on the molecular basis of the causes of lung cancer, hoping to sow doubt about the dangers of smoking in the larger population.

The defendants in some toxic tort cases are looking at the genetic predisposition of certain plaintiffs as a way to shift blame. In a products liability suit, for example, a plaintiff blamed his birth defect on *in utero* exposure to toxins at the plant where his mother worked. However, the company owning the plant claimed in its defense that a genetic disorder caused the defects, not exposure to toxins.[37] In a similar liability case, a company tried to compel a plaintiff to take a test for Fragile-X Syndrome, insisting that his disability was not caused by toxic substances, but was innate.[38]

Conclusion

Molecular biologists have justified their search for genes and genetic markers through claims about the diagnostic value of genetic information for the individuals who are predisposed to a genetic disease. There are some who question their claims, for at present, diagnostic predictions have few therapeutic implications for the person at risk. Extending genetic predictions to behavior, however, may have important implications even when based less on science than on belief. The concept of predisposition, I have argued, is framing the way we think about individual success or failure and about the sources of social problems. Yet the deterministic assumptions underlying the concept, especially when applied to complex behaviors such as violence, are highly questionable. As neuroscientist Steven Rose writes, the tendency towards neurogenetic determinism is based on "a faulty reductive sequence whose steps include: reification, arbitrary agglomeration, improper quantification,... misplaced causality, and dichotomous partitioning between genetic and environmental causes."[39] However, despite the complexity of behavior, the limited understanding of causality, and the poorly understood influence of social and environmental factors, biological theories and predictions are gaining importance in the implementation and justification of social policies.

As a growing range of behaviors and conditions are being defined in terms of genetics, the dilemmas can be profound. A person who is asymptomatic, but identified as predisposed to a disease or to

---

[36] Michael Miller, "In Vino Veritas," *Wall Street Journal*, June 8, 1994.

[37] Severson v. Markem Corp., No. 698517, Feb 26, 1990.

[38] Paul Billings, personal communication, July 17, 1996.

[39] Steven Rose, "The Rise of Genetic Determinism," *Nature*, 372, 1995. 380-382.

anti-social behavior, may lose insurance or employment as the statistically driven concept is interpreted as a definitive prediction. Children identified as predisposed to violence can be medicalized, labeled, and stigmatized. And the fatalistic idea of genetic predisposition can also encourage passive attitudes toward social injustice and apathy about continuing social problems—especially in societies preoccupied with cost containment in the social policy arena.

In this paper, I have repeatedly pointed to historical parallels to suggest the persistent appeal of explanations based on the idea of predisposition. Such explanations provide supposedly science-based reasons to justify social policies and to preserve the status quo. As a source of predictive information, science-based ideas about biological predispositions are seductive. They permit institutions to abdicate responsibility by blaming the individual—the person predisposed to violence or addiction, the individual genetically flawed. Much like the wisdom of shamans, soothsayers, and others who have tried to predict the future, the predictive claims of genetics can serve ideological agendas. They are a way to protect existing social categories and social policies while promising control of those who are defined as a threat to the social order.

SCALE, AWARENESS AND CONSCIENCE: THE MORAL TERRAIN OF ECOLOGICAL VULNERABILITY

Robert H. Socolow

Prosperity is stressing the environment. This interaction can be illuminated by separating out aggregate size (*scale*), available science (*awareness*), and the obligation to respond (*conscience*). Solutions require an evolving vision of the good life, a sustained commitment to open science, and both an active and a reverent management of the Earth.

The three-dimensionality of ecological vulnerability

Ecological vulnerability is a big new idea, far more prominent on the intellectual landscape now than thirty years ago. At the core of this idea is the observation that our exuberant species is transforming the Earth's natural systems. Examples are both regional and global. Regional impacts include depleted fisheries and unhealthy urban air. Global impacts include less ozone in the stratosphere, a more acidic surface ocean, and higher atmospheric concentrations of carbon dioxide, methane, and several other greenhouse gases.

Ecological vulnerability is a permanent, if not yet familiar, feature of the human condition, a new element of the human tragedy. Why has it become a prominent concern late in the twentieth century? There are three contributing factors:

1. The magnitudes of our effects (scale)
2. Our capacity to understand these effects (awareness)
3. The obligation we feel to respond to these effects (conscience)

Let us take these three factors as the three dimensions of ecological vulnerability. The salience of ecological vulnerability in our time—its substantiality, its obtrusive presence—is the consequence of the maturing of contemporary reality along all three of these dimensions. We humans are confronting a new world where our prosperity, our awareness of the consequences of our prosperity, and our determination to address these consequences have all risen above some threshold of significance at approximately the same time. The three-dimensionality of ecological vulnerability provides helpful coordinates for exploring its moral terrain.

Two-dimensional variants

We can easily imagine worlds where only two of the three dimensions of ecological vulnerability have matured, and one is nearly absent. It is instructive to identify each of these worlds.

2&3,but not 1: Awareness and conscience have matured, but not scale. Such an imaginary planet is much larger than ours, with the result that, in all instances, the impacts on environmental systems resulting from the actions of its human population are dwarfed by the impacts of natural processes. The relationship between cumulative human impact and natural capacity resembles what was once found (say, in the eighteenth century) on Earth. On the imaginary planet, however, environmental science is just as well developed as ours today, capable of measuring and modeling both natural systems in the absence of human activity and the impacts of human activity on natural systems. And the moral sensibilities of the citizens are finely honed.

Accordingly, people in this imaginary world struggle to find appropriate ways to express their concern for future generations, frustrated that their science is telling them their efforts will make hardly any difference. Wishing the message to be false, they spend their time carefully documenting the minutiae of human activity. They are punctilious in the observance of ecological correctness. Theirs is a world of fetishism and misplaced angst.

People in our (real, current) world sometimes assert that this imaginary world properly describes life on Earth today. They insist that our human society is still too puny to overwhelm nature. For example, in spite of abundant evidence to the contrary, they are captivated by arguments that the depletion of the stratospheric ozone layer is the result of gases coming from volcanoes; in their model of reality, the causative agent cannot possibly be man-made chlorofluorocarbons.

1&3, but not 2: Scale and conscience have matured, but not awareness. In this second imaginary world, impacts are indeed substantial, and people are morally sensitive about imposing risks on others. But there is no science available to guide action, and, as a result, people in this imaginary world fly blind. Resembling this imaginary world, perhaps, was our planet in the years of the Black Death in the fourteenth century, when bubonic plague was spreading relentlessly, and there was little understanding of infectious disease.

People in our own present world sometimes assert that this second imaginary world properly describes life on Earth today. They focus on the deficiencies in our environmentally relevant science, rather than on our prowess. At every juncture, their preference is for delay. They emphasize that we are not yet in full command of the facts and that if we wait until we have more information, we will find better ways out of trouble.

1&2, but not 3: Scale and awareness have matured, but not conscience. In this third of our three imaginary worlds, effects of human action are large and understood to be large, but attention is elsewhere. Of concern is only the present day: *après nous le déluge*. Ethics are tribal. Crises arise, preventive action has not been taken, and damage is high. Many suffer, the weak especially.

People in our own world usually do not assert that this imaginary world describes life on Earth. No one wants to concede that we are inadequately endowed with conscience. But many argue that our obligations to future generations are minimal, relative to our obligations to those closest to us today, including ourselves. Their priorities resemble the priorities in this third, imaginary world.

We are not strangers in any of these worlds. In fact, our behavior is often more appropriate for one of these worlds than for our own. Each is a convenient refuge from reality.

An example: Human impacts on the nitrogen cycle

We find examples of ecological vulnerability almost everywhere today. Again and again, a productive line of inquiry is to ask how some particular natural system works and what effects human activity is having upon it. One finds, consistently, that at least some effects are large, that the significance of these effects is partially (but not fully) understood, and that actions designed to reduce these effects are ethically complex.

Such a pattern of answers is found today when one inquires about the natural cycles of many, if not most, of the elements in the periodic table. Consider the nitrogen cycle. In the overall global nitrogen cycle, there are two subcycles: a fixing-unfixing subcycle and a fixed-nitrogen cycle. In the fixing-unfixing subcycle, certain species of "nitrogen-fixing" bacteria change atmospheric nitrogen (N_2) into forms usable by plants (simple single-nitrogen molecules, called "nutrients"); simultaneously, other, "denitrifying" (in effect, unfixing) bacteria return the nitrogen to the atmosphere by rebuilding N_2 from nitrogen nutrients. In the fixed-nitrogen cycle, fixed nitrogen passes through many fixed chemical forms: nutrients are taken up by living plants and animals, while simultaneously the nitrogen in plants and animals, usually long after their death, is returned to nutrient forms through decomposition. Once within the fixed-nitrogen subcycle, a nitrogen atom cycles back and forth between nutrient and plant matter many times before it is recycled to the atmosphere through the fixing-unfixing subcycle.[1]

Inspecting the ancient air archived in ice cores in Arctic and Antarctic glaciers, scientists can infer that between two thousand ago and as recently as fifty years ago the overall nitrogen cycle was in approximate equilibrium. In particular, through the fixing-unfixing cycle, nitrogen-fixing bacteria transformed atmospheric nitrogen into nitrogen nutrient at a rate of between 100 and 200 million metric tons of nitrogen

---

[1] Ayres, Schlesinger, and Socolow 1994, Kinzig and Socolow 1994.

per year (100-200 Mt(N)/yr), and denitrifying bacteria transformed nitrogen in nutrients back to atmospheric nitrogen at approximately the same rate.

Human activity is currently doubling the previous rate of nitrogen fixation. The largest contributor is nitrogen fertilizer production, which begins with the production of ammonia from atmospheric nitrogen. Driven by the Green Revolution, global nitrogen fertilizer production has grown more than six-fold in the past 35 years: from 13 Mt(N)/yr in 1961 to 87 Mt(N)/yr in 1996.[2] The two other large contributors to anthropogenic nitrogen fixation are deliberate planting of nitrogen-fixing crops, such as soybean and alfalfa (roughly, 40 Mt(N)/yr) and high-temperature combustion in engines and boilers (roughly, 30 Mt(N)/yr).[3]

The *scale* of human impact on a natural phenomenon can be measured as a fraction, with an effect in the absence of human activity in the denominator and the incremental effect of human activity in the numerator. When the fraction captures an important environmental concern, and the value of the fraction is close to unity (1.0) or larger, it is reasonable to call the effect "large." Thus, human impact on the nitrogen cycle, measured by fixed-nitrogen production, meets the test of large scale.

What about *awareness*? Both medicine and environmental science contribute insights into the damage caused by the build-up of fixed nitrogen in water and air.[4] An excess of fixed nitrogen can contaminate drinking water and can contribute to unhealthy air; it can also intensify acid deposition, eutrophication, stratospheric ozone depletion, and the greenhouse effect. Cumulative impact, in most cases, will depend on whether denitrifying bacteria keep pace with incremental nitrogen fixation, thereby preventing a relentless build-up of regional and global stocks of fixed nitrogen. Assuming a continuous build-up of fixed nitrogen does occur, the most troublesome long-term impact may be the disruption of ecological balance in unmanaged natural areas, as indiscriminate fertilization brings about differential responses across species, diminishing biodiversity and degrading ecosystem services.[5] Because air and water can transport excess fixed nitrogen large distances, sites of inadvertent fertilization will often be far from the sites of deliberate fertilization for agriculture.

What about *conscience*? The environmental costs of excess fixed-nitrogen are a challenge to moral reasoning. The Green Revolution's high-yield agriculture has nearly permitted global food

---

[2] International Fertilizer Industry Association 1998.
[3] Galloway 1998.
[4] Socolow 1999.
[5] Tilman 1982, Schlesinger 1994, Vitousek 1994, Vitousek et al. 1997.

production to keep pace with global population growth over the past three decades. Because nitrogen fertilizer has made possible these high yields, it is sometimes regarded as a gift from the gods. Idolizing a technology or physical resource, however, is never wise. Nitrogen fertilizer is just a commodity, much of it used to meet the food preferences of the prosperous.

After absorbing the messages of scale, awareness, and conscience, people must act. One goal is "precision agriculture," which fine-tunes timing and quantity of fertilizer application. Other goals include improving the management of crop residues, reintegrating grain agriculture and animal husbandry, and reducing waste in food transport and storage. Targeted subsidies can address the needs of the poor, and market mechanisms such as trading in fixed-nitrogen emissions at various spatial scales can help identify high-payoff environmental investments.[6]

The above analysis of the fixed-nitrogen issue is prototypical. Considerations of scale, awareness, and conscience triangulate every environmental issue. For the remainder of this essay we examine how scale, awareness and conscience, illuminate aspects of environmental ethics.

The ethical content of the dimension of scale

Consider, first, *the ethical content of scale*. It is here that we encounter judgments about consumption. The scale of environmental impact is directly related to the level of consumption measured in physical (but often not economic) units. When consumption is viewed in physical terms, it is associated with flows of materials: their extraction from below ground, from water, or from the air; their chemical and physical transformation; and their return to the environment, generally in altered form.

A physical view of consumption does not penetrate to the core of the phenomenon. Working inward, the next layer of analysis probes why a particular desired human benefit involves particular resource impacts (why lighting is provided by certain lightbulbs, why visiting a friend is accomplished with a particular vehicle); the analysis probes choices among available technologies.

At a still deeper level one asks why particular benefits are desired. Here, one encounters territory that is surprisingly little explored. Identifying the consumption activities that are environmentally significant (travel by vehicle, for example) has been a far more active area of research than probing the determinants of consumption (why travel is desired).[7]

---

[6] Socolow 1999.
[7] United Nations Secretariat 1997, Stern et al. 1997.

The dominant model of the determinants of consumption is the ladder. The ladder has a sequence of rungs. Those on each rung have an income somewhat lower than the income on the rung above. With the passage of time, a household with rising income will tend to climb the rungs, adopting the consumption patterns of each higher rung in turn. The driver of consumption, by this model, is emulation. Such a ladder model has been confirmed in studies of fuel choice for cooking, for example. With increasing income in many developing countries, cooking fuel changes, predictably, from firewood to charcoal to liquid petroleum gas to natural gas and electricity.[8]

Even the most simplified ladder model does not presume that the per capita environmental impact of consumption simply increases as the ladder is climbed. Environmental damage from the consumption of the very poor often exceeds environmental damage from consumption a few rungs above. Accelerated slash-and-burn agriculture that does not give the land time to recover, cooking with wood on an open fire, discharging untreated wastes into rivers that become sewers—these are some well-known examples of poverty's resource-intensiveness, its disproportionate impact on the environment. Higher up the ladder, however, environmental impact increases with income.

Population size enters the ladder model only as a multiplicative factor, and thus tends to disappear from view. Yet population size is obviously critical to impacts due to scale. And population size has direct impacts on individual expenditures that a multiplicative model cannot capture. Strains produced by total number of people engender expenditures on pollution control and expenditures to obtain privacy.[9]

The ladder model, in its most elementary form, presumes that there is just one ladder and that the ladder has no branches. From the nomad to the sheik there is a single ascent. From the urban slum dweller to the denizen of a suburban estate there is also a single ascent. The model of an unbranched ladder is gaining support today, as an increasingly large fraction of the world's population assents to a single concept of the good life. For shorthand, call this concept "Western." The good life is equated with the maximization of experiences. It is marked by an emphasis on self-realization, a zest for activity, a search for comfort and autonomy, and an accumulation of goods. Travel prevails over sitting still, and cars over buses.

For much of history, civilizations, especially in Asia, defined the good life differently. Self-examination, inner peace, subordination of desire – all were advocated effectively. Serious people once questioned whether the "undeveloped" countries, including those in Asia, could develop along Western lines, or even wished to. In the recent past,

---

[8] Sathaye and Tyler 1991.
[9] Cohen 1995; Kates 1997, Bloom and Williamson 1997.

however, "undeveloped" was replaced by "underdeveloped," or "less developed," or "developing." There are no separate categories for "conventionally developing" and "unconventionally developing." The issue, as far as the West is concerned, is settled. Apart from a few special cases where traditional alternatives are still being explored, the western ideal of the good life is unchallenged globally today. Development is no longer a matter of "whether," nor of "how," but only of "when."

People wish to use the ladder model to discern the environmental consequences of future consumption. A ladder model with static preferences at every rung predicts that aggregate environmental damage from human consumption will increase, once the dominant ascent on the ladder is not from very poor to poor but from poor to what is loosely called "middle class." An amended ladder model that takes into account the effects on preferences of technological change provides a more ambiguous forecast, because, in many instances, the direction of technological change is toward reduced environmental damage per unit of consumption. As a result, greater environmental damage is not an inevitable accompaniment to increased global economic well-being. Nonetheless, there are ample grounds for concern that the planet's environment may be significantly degraded, even with technologies designed to reduce environmental impacts, if ten billion people reach today's higher rungs.

In addition to technological change, there will be changes in values. The ladder model at its most simplistic locates these changes only on the highest rung, where there is no one to emulate. What is judged desirable by the most privileged people is endowed with glamour a few rungs below and is copied in each successive generation by ever larger numbers of people. The long-term environmental consequences of the American heavy automobile are highlighted by such a model.

In fact, however, value change originates at every rung of income or social status. It is not decreed from the top. The decisions to spend substantial private and public resources to clean the air and water and to protect land from development have been broadly based consumption decisions. Even more broadly based has been the world-wide decision to have fewer children. Future regional and even global average fertility rates below replacement level begin to be credible. When such rates persist, total population does not stay on a plateau after reaching its peak (as embedded in all orthodox projections), but decreases indefinitely. In decisions about pollution, land, and child-bearing one can discern an adaptive feedback that may substantially transform the environmental prospect.

The ladder model has a tight grip on today's visions of the future. Accordingly, it needs careful testing. Especially worthwhile are explorations of its key assumption that consumer choice is dominated by emulation of the choices made on the rungs immediately above. It is also critical to gain insight into how consumption patterns change to reflect evolving values. From an environmental perspective, several trends are positive.[10]

The ethical content of the dimension of awareness

Consider, next, *the ethical content of awareness*. It is here that we encounter judgments about science and technology. There are two imperatives: 1) to enhance those portions of the scientific enterprise likely to illuminate critical environmental issues, and 2) to strengthen the norms of science in arenas of controversy.

Wherein lies the moral imperative to enhance those portions of the scientific enterprise likely to illuminate critical environmental issues? It arises from our obligation to preserve the capacity of future generations to enjoy experiences that they value as much as we enjoy what we value. Such a formulation was made explicit in the report of a United Nations commission, *Our Common Future*, popularly known as the Brundtland Report, after Gro Harlem Brundtland, the commission's chairman: "Humanity has the ability to make development sustainable – to ensure that it meets the needs of the present without compromising the ability of future generations to meet their own needs."[11]

To achieve such an objective, each generation must provide the next generation with new capabilities in order to compensate for bequeathing to the next generation a natural environment more degraded than the one it inherited. Such degradation will not be universal, but it is likely to be pervasive. The Earth's largest and purest or highest-grade (lowest entropy) oilfields, fresh-water aquifers, and mineral deposits are likely to disappear, for example, so that future generations will have to make do with only the Earth's next lowest entropy resources. Where geology threatens to impoverish, the intergenerational accounts can be balanced by scientific understanding, new instruments and devices, and more subtle and effective policies. An obligation to provide future generations with enhanced capabilities is easier to defend than an obligation to preserve every particular experience available today.[12] To be sure, we retain specific responsibilities for unique features of our natural, aesthetic, and historical legacy. And we have obligations toward species other than ourselves.

Enhanced capabilities require the vigorous pursuit of what the late Donald Stokes, in *Pasteur's Quadrant*, calls "use-inspired basic research,"[13] research that both addresses practical concerns (here, environmental compatibility) and searches for generalizable principles.

---

[10] Kempton, Boster, and Hartley 1995, Daly 1996.
[11] World Commission on Environment and Development 1987, p.8.
[12] Solow 1991.
[13] Stokes 1997.

Two subjects of contemporary use-inspired basic research are ecosystem services and industrial ecology. Ecosystem services are the benefits human beings gain from natural ecosystems, ranging from cleansing the air and water to enabling such finely tuned processes as pollination. Research on ecosystems is focusing, for example, on the relationship between alteration of the species composition of an ecosystem and degradation of its service functions.[14]

Industrial ecology focuses on how the industrial system is embedded in the natural environment.[15] Firms and networks of firms are center stage. Flows of materials through the economy are visualized as rearrangements of the stuff we find on or close to the surface of the Earth.[16] Research in industrial ecology focuses, for example, on the recycling and scrap industries (which "close loops," extending the time materials spend within the industrial system), and on patterns of ownership involving lease rather than sale. Among the systems recently studied are lead,[17] wood products,[18] metals,[19] and nitrogen.[20]

Just as important as deepening the pursuit of use-inspired basic research is strengthening the norms of science in arenas of controversy. Science gains its strength from 1) the methodology of trial and error, 2) the tradition of open access to results, and 3) the presumption of international collaboration.

The methodology of trial and error strengthens science by not condemning failure. Research is pursued along parallel tracks, with the expectation that many tracks will later be abandoned. It has been said that in order to know the truth, it is necessary to imagine a thousand falsehoods. Deciding which research areas to pursue more deeply and which approaches to replicate widely is an iterative process, informed by the evaluation of each completed experiment.

The tradition of open access to results strengthens science by providing a mechanism for peer review, replication, error detection, and the transfer of knowledge from one application to another. Furthermore, open access is the engine of democratization. It encourages debate and the formation of consensus. Open access permits media involvement, at its best fostering a comprehending and properly skeptical public.

The presumption of international collaboration strengthens science by permitting any scientist or engineer to address any problem, irrespective of where the individual works and where the problem is

---

[14] Daily 1997.
[15] Socolow et al. 1994; Graedel and Allenby 1994, Ayres and Simonis 1994, Frosch 1997.
[16] Socolow 1994.
[17] Socolow and Thomas 1997.
[18] Wernick, Waggoner, and Ausubel 1998.
[19] Wernick and Themelis 1998.
[20] Socolow 1999.

particularly pressing. Problems judged to have the greatest prestige or market interest gain broad-based international attention. Unfortunately, these are usually not the most pressing environmental problems.

The ethical content of the dimension of conscience

Consider, finally, *the ethical content of conscience*. It is here that we encounter judgments about deliberate action. Two kinds of deliberate action are often distinguished: mitigation and adaptation. Mitigation is preventative; it is designed to reduce the magnitudes of expected effects or their rate of onset. Mitigation includes both "first-generation" end-of-pipe pollution control and, increasingly, "next generation" clean processes that embed ecological concerns.[21] Adaptation is designed to reduce the consequences of the expected effects after they arrive. Adopting energy efficient technology is mitigation; building dikes is adaptation. Some actions are a combination: planting a forest may both sequester atmospheric carbon (mitigation) and reduce the hydrological and ecological impacts of future climate change (adaptation).

The ethicist may ask whom mitigation or adaptation is intended to benefit. Investments addressing ecological vulnerability can be compared with other self-interested investments on behalf of the next one or two generations, such as investments in education. It is not easy to assess whether, within and across nations, a particular package of mitigation and adaptation investments is progressive or regressive. The beneficiaries are likely to include both those whose basic needs are at risk and those who have a large financial stake. A mitigation program to slow the rise in sea level helps both the millions of poor people who live in the floodplain of the Ganges in Bangladesh and those with million-dollar homes at a beach.

Investments on behalf of even the poorest members of a future generation can be regressive, relative to investments on behalf of today's poor.[22] Targeted investments on behalf of today's poor are morally compelling. A world is within reach where the entire population is adequately fed, has access to safe drinking water, and is free of many infectious diseases. The greater the success in providing basic human needs for today's people, the more compelling the case for addressing the ecological vulnerability of future generations.

A mitigation program can be judged by whether it can proceed in small steps that permit incremental learning, and by whether there is room for turning back. Such judgments are particularly germane to "Earth systems engineering" proposals, which are global in scope.[23] An example is the proposal to continue the use of fossil fuels while

---

[21] Chertow and Esty 1997.
[22] Schelling 1992.
[23] Keith and Dowlatabadi 1992, Allenby 1998.

"sequestering" the resultant carbon dioxide so that it does not enter the atmosphere, thereby slowing the rate of climate change. Fossil fuel is chemically processed into hydrogen fuel and byproduct carbon dioxide, and the carbon dioxide is piped into deep aquifers or the deep ocean. The merits of this proposal are being widely discussed.[24]

The processes by which Earth systems engineering proposals are evaluated require legitimating. Some cures are worse than the disease. And some cures will work. There are many parallels with the engineering of the genome. Who will guard the guardians? And who will protect us from guardians of the guardians who see their task as avoiding every potentially slippery slope, thereby annulling the spirit of experimentation, so critical to our future?

Finally, how can the spirit of experimentation be reconciled with reverence? We are changing the Earth too much and too quickly to be able to defend doing nothing.[25] But the Earth is sacred. It needs not only our stewardship, but our love.

Toward more persuasive prophesy

The prophets of ecological vulnerability would like to be as persuasive as their contemporaries (often the same people) who preach the need to prevent nuclear war. Nuclear prophets articulate the terrible costs of wrong action, and they are believed. Ecological prophets use a similar vocabulary. But their foresight is doubted. Their message is resisted.

The difference is not only that ecological prophets have arrived more recently on the scene. (After all, the message of the nuclear prophets was resisted too, for several decades.) The difference is also the consequence of the greater incompleteness of ecological prophesy.

Help can come from all three of the dimensions of ecological vulnerability. If the *scale* of human impact can grow less quickly, as a result of first-round interventions like a global climate agreement, valuable time will become available. That time can be used to improve *awareness*: targeted experiments in science, technology and policy can clarify the more and less serious environmental threats and the more and less promising responses. And, in that same time, our *consciences* can be more fully engaged in understanding what is required to manage the Earth actively, yet with reverence.

Acknowledgement

The invitations of Arthur Galston to speak in his lecture series and to write for this volume build upon a relationship that began almost thirty years ago. At that time I was

---

[24] Herzog 1997, Hileman 1997, Socolow 1997, Parson and Keith 1998.
[25] Dyson 1975.

76

searching for ways to bring my disciplinary background in pure physics to bear on the relatively unstructured problems in environmental studies. He was one of the few senior professors at Yale (where I was an Assistant Professor) who said it could be done. He was one of an even smaller number who said it *should* be done. I gave it a try, and I am still at it. I am hugely in Arthur's debt for his encouragement and vision.

References

Allenby, Braden, 1998. "The role of industrial ecology in an engineered world." To be published in *Journal of Industrial Ecology*, MIT Press.

Ayres, Robert, William Schlesinger, and Robert Socolow, 1994. "Human impacts on the carbon and nitrogen cycles." In Industrial Ecology and Global Change, R. Socolow, C. Andrews, F. Berkhout, and V. Thomas, eds. Cambridge, UK: Cambridge University Press. 121-155.

Ayres, Robert, and Udo Simonis, eds., 1994. Industrial Metabolism: Restructuring for Sustainable Development. Tokyo: United Nations University Press.

Bloom, David, and Jeffrey Williamson, 1997. "Demographic transitions and economic miracles in emerging Asia." *National Bureau of Economic Research Working Paper 6268*. Cambridge, MA: National Bureau of Economic Research.

Chertow, Marian, and Daniel Esty, eds., 1997. Thinking Ecologically: The Next Generation of Environmental Policy. New Haven, CT: Yale University Press.

Cohen, Joel, 1995. How Many People Can the Earth Support? New York: Norton & Co.

Daily, Gretchen, ed., 1997. Nature's Services: Societal Dependence on Natural Ecosystems. Washington, D.C.: Island Press.

Daly, Herman, 1996. Beyond Growth. Boston, MA: Beacon Press.

Dyson, Freeman, 197. "The hidden cost of saying No!" *Bulletin of the Atomic Scientists*, June 1975, 23-27.

Frosch, Robert, 1997. "Toward the end of waste: Reflections on a new ecology of industry." In Technological Trajectories and the Human Environment, Jesse Ausubel and Dale Langford, eds. Washington, D.C.: National Academy Press, 157-167.

Galloway, James, 1998. "The global nitrogen cycle: changes and consequences." *Environmental Pollution*, in press.

Graedel, Thomas, and Braden Allenby, 1994. Industrial Ecology. Upper Saddle River, NJ: Prentice Hall.

Herzog, H.J., ed., 1997. Proceedings of the Third International Conference on Carbon Dioxide Removal, Cambridge, MA, 9-11 September 1996, *Energy Conversion and Management*, Vol. 38, Suppl.

Hileman, Bette, 1997. "Fossil Fuels in a Greenhouse World." *Chemical and Engineering News*, 75:34-37, August 18, 1997.

International Fertilizer Industry Association, 1998. Webpage at http://www.fertilizer.org.

Kates, Robert, 1997. "Population, technology, and the human environment: A thread through time." In Technological Trajectories and the Human Environment, Jesse Ausubel and Dale Langford, eds. Washington, D.C.: National Academy Press, 33-55.

Keith, David W., and Hadi Dowlatabadi, 1992. "A serious look at geoengineering." *Eos* 73(27), 289-296.

Kempton, Willett, James Boster, and Jennifer Hartley, 1995. Environmental Values in American Culture. Cambridge, MA: MIT Press.

Kinzig, Ann, and Robert Socolow, 1994. "Human impacts on the nitrogen cycle." *Physics Today* 24(11):24-31.

Parson, E.A., and D.W. Keith, 1998. "Fossil fuels without CO2 emissions" *Science*, Vol 282, No 5391, 6 November 1998, 1053-1054.

Sathaye, Jayant, and Stephen Tyler, 1991. "Transitions in household energy use in urban China, India, the Philippines, Thailand, and Hong Kong." *Annual Review of Energy and the Environment, 1991*, 16:295-335.

Schelling, Thomas, 1992, "Some economics of global warming." *American Economic Review*, May 1992. Reprinted in Economics of the Environment: Selected Readings, 3rd edition, Robert Dorfman and Nancy S. Dorfman, eds. New York: Norton & Co.,1993, 464-483.

Schlesinger, W.H., 1994. "The vulnerability of biotic diversity." In Industrial Ecology and Global Change, R. Socolow, C. Andrews, F. Berkhout, and V. Thomas, eds. Cambridge, UK: Cambridge University Press, 245-260.

Socolow, Robert, Clinton Andrews, Frans Berkhout, and Valerie Thomas, eds., 1994. *Industrial Ecology and Global Change*. Cambridge, UK: Cambridge University Press.

Socolow, Robert, 1994. "Six perspectives from industrial ecology." In Industrial Ecology and Global Change, R. Socolow, C. Andrews, F. Berkhout, and V. Thomas, eds. Cambridge, UK: Cambridge University Press. 3-16.

Socolow, Robert, and Valerie Thomas, 1997. "The industrial ecology of lead and electric vehicles." *Journal of Industrial Ecology* 1(1):13-36.

Socolow, Robert , ed. 1997. Fuels Decarbonization and Carbon Sequestration: Report of a Workshop. Princeton, NJ: Princeton University, PU-CEES Report 302, September 1997.

Socolow, Robert, 1999. "Nitrogen management and the future of food: Lessons from the management of energy and carbon." *Proceedings of the National Academy of Sciences* (to be published).

Solow, Robert M. 1991. "Sustainability: An Economist's Perspective." Eighteenth J. Seward Johnson Lecture to the Marine Policy Center, Woods Hole Oceanographic Institution, Woods Hole MA June 14, 1991. Reprinted in Economics of the Environment: Selected Readings, 3rd edition. Robert Dorfman and Nancy S. Dorfman, eds. New York: Norton & Co.,1993, 179-187.

Stern, Paul, Thomas Dietz, Vernon Ruttan, Robert Socolow, James Sweeney, eds, 1997. Environmentally Significant Consumption: Research Directions. Washington, D.C.: National Academy Press.

Stokes, Donald E., 1997. Pasteur's Quadrant: Basic Science and Technological Innovation. Washington, D.C., Brookings Institution Press.

Tilman, D. 1982. Resource Competition and Community Structure. Princeton University Press, Princeton, New Jersey.

United Nations Secretariat, 1997. Critical Trends: Global Change and Sustainable Development. New York: United Nations Department for Policy Coordination and Sustainable Development.

Vitousek, P. M. 1994. "Beyond global warming: ecology and global change." *Ecology* 75:1861-1876.

Vitousek, P. M., J. D. Aber, R. W. Howarth, G. E. Likens, P. A. Matson, D. W. Schindler, W. H. Schlesinger and D. Tilman. 1997. "Human alteration of the global nitrogen cycle: sources and consequences." *Ecological Applications* 7(3):737-750.

Wernick, Iddo, Paul Waggoner, and Jesse Ausubel, 1998. "Searching for leverage to conserve forests: The industrial ecology of wood products in the United States." *Journal of Industrial Ecology*, 1(3):125-145.

Wernick, Iddo, and Nickolas Themelis, 1998. "Recycling metals for the environment." *Annual Review of Energy and the Environment, 1998* 23:465-497.

World Commission on Environment and Development, 1987. Our Common Future. New York: Oxford University Press.

PESTICIDE USE: ETHICAL, ENVIRONMENTAL, AND PUBLIC HEALTH IMPLICATIONS

David Pimentel and Kelsey Hart

Introduction

Worldwide, over 2.5 million tons of pesticides are applied each year[1] at a purchase price of $32 billion. In the United States alone, approximately 500,000 tons of 600 different types of pesticides are used annually,[2] with a yearly price tag of $6.5 billion, including application costs.[3]

Despite widespread use of pesticides in the United States, pests (principally insects, plant pathogens, and weeds) destroy 37% of all potential food and fiber crops.[4] Estimates are that losses to pests would increase by 10% if no pesticides were used at all; specific crop losses would range from zero to nearly 100%.[5] Thus, pesticides make a significant contribution to maintaining world food production. In general, each dollar invested in pesticide control returns about $4 in crops saved.[6]

Although pesticides are generally economically profitable, their use alone does not always decrease crop losses. For example, even with the tenfold increase in insecticide use in the United States from 1945 to 1989, total crop losses from insect damage have nearly doubled from 7% to 13%.[7] The recent abandonment of some sound agricultural practices—crop rotations, for instance—in many parts of the country is primarily responsible for these increased losses.

Most of the commonly cited benefits of pesticides are based only on direct crop returns; however, such assessments do not consider the indirect but substantial environmental, economic and social costs associated with pesticide use. It has been estimated that only 0.1% of applied pesticides reach the target pests, while the bulk of each pesticide application (99.9%) is left to impact the surrounding environment.[8] These impacts must be closely examined to facilitate the development and implementation of a balanced, sound policy of pesticide use.

The obvious need for current and comprehensive study of pesticide impact prompted an investigation by Pimentel et al. (1993) of the complex environmental and economic costs resulting from the

---

[1] PN, 1990.
[2] Pimentel et al., 1991.
[3] USBC, 1994.
[4] Pimentel et al., 1991.
[5] Pimentel et al., 1978.
[6] Pimentel et al., 1991.
[7] Id.
[8] Pimentel 1995.

nation's dependence on pesticides. Included in that assessment are analyses of pesticide impacts on human health; domestic animal poisonings; increased control expenses resulting from pesticide-related destruction of natural enemies and from the development of pesticide resistance; crop pollination problems and honey bee losses; crop and crop product losses; groundwater and surface water contamination; fish, wildlife, and microorganism losses; and governmental expenditures to reduce the environmental and social costs of pesticide use. Parts of this paper are condensed from that work.

With the publication of Rachel Carson's Silent Spring in 1962, the general societal attitude towards the widespread use of pesticides and other chemicals has become highly critical. The significant environmental and health impacts raise important ethical questions. Careful consideration of the often adverse environmental and social effects of pesticides, coupled with their overall economic benefits, leads us to ask whether using synthetic pesticides is ethically acceptable. If it can be shown that pesticides cause widespread and long-lasting adverse effects, is it morally responsible to continue using them? Or would ceasing the use of synthetic pesticides actually cause more economic, and in turn social, damage to the world and its people? Balancing the diverse costs and benefits related to pesticide use is an extremely complex undertaking.

In this paper, through a careful analysis of the environmental, social and economic impacts of pesticides, we address some difficult ethical concerns pertinent to pesticide use.

Pesticide poisonings and public health concerns
Since the first use of DDT for crop protection in 1945, the total amount of pesticides used in agriculture worldwide has been staggering. In 1945, about 50 million kg of pesticides were applied worldwide. Today, global usage is at about 2.5 billion kg per year, an approximate fifty-fold increase since 1945.[9] Unfortunately, the toxicity of most modern pesticides is more than tenfold greater than those used in the early 1950s,[10] so the potential hazards have increased as well.

In 1945, when synthetic pesticides were first used, there were few reported pesticide poisonings. But by the late 1960s, both pesticide usage and toxicity had increased so dramatically that the number of human pesticide poisonings was substantial.[11] Unfortunately, this trend has continued into the present. Just in the last decade, the total number of pesticide poisonings in the United States has increased from 67,000 in

---

[9] Id.
[10] Id.
[11] Id.

1989 to the current level of 110,000 per year.[12] Worldwide, the increased use of pesticides results in approximately 26.5 million cases of occupational pesticide poisonings each year, and an unknown number of non-occupational pesticide poisonings.[13] Of all these estimated poisoning episodes, about 3 million cases are hospitalized, resulting in approximately 220,000 fatalities and about 750,000 cases of chronic illness every year.[14]

Chronic effects of pesticides are diverse and adversely affect most systems of the human body. Schuman[15] lists six types of health effects that pesticide over-exposure can result in: acute toxicity; sub-acute delayed toxicity; chronic cumulative toxicity; reproductive effects; hypersensitivity; and psychological conditioning to time/place exposures. Specific effects can range from death to chronic illness to exposure-induced chemical sensitivities and allergies. U.S. data indicate that 18% of all insecticides, and about 90% of all fungicides, are carcinogenic,[16] and many studies have shown that risks for certain types of cancers are higher in people—like farm workers and insecticide applicators—who are more frequently exposed to certain pesticides.[17] Many pesticides are also estrogenic, perhaps suggesting a linkage to the increased breast cancer rate among some groups of women in the U.S. The breast cancer rate rose from 1 in 20 in 1960 to 1 in 8 in 1995;[18] there was a concurrent increase in pesticide use during that period, although it has not been concretely linked to the increase in breast cancer rates.

In addition to their carcinogenic effects, pesticides can have adverse effects on the respiratory and reproductive systems. For example, 15% of a group of professional pesticide applicators suffered asthma, chronic bronchitis, or chronic sinusitis, as compared with 2% of people who used pesticides infrequently.[19] Studies have also linked pesticides with reproductive effects such as infertility and fetal deformities, but the data are still inconclusive.

The negative health effects that pesticides can have are more significant in children, for several important reasons. First, children have much higher metabolic rates than adults, and their ability to activate, detoxify, and excrete toxic compounds is different from that of adults. Also, because of their smaller physical size, children are exposed to higher levels of pesticides per unit of body weight. In addition, certain types of pesticides are more dangerous for children than for adults. For

---

[12] Litovitz et al. 1990; Benbrook et al. 1996.
[13] UNEP 1997.
[14] WHO 1992.
[15] Schuman 1993.
[16] NAS 1987.
[17] Culliney et al. 1993.
[18] McCarthy 1993.
[19] Weiner and Worth 1972.

example, the organophosphate and carbamate classes of pesticides adversely affect the nervous system by inhibiting cholinesterase, a critical enzyme. This problem is particularly significant for children since their brains are more than five times larger in proportion to their body weight than those of adults. In a California study, 40% of the children working in agricultural fields had blood cholinesterase levels below normal, a strong indication of organophosphate and carbamate pesticide poisoning.[20]

With the increased understanding of the distinct physiological differences between adults and children, it has become obvious that the present pesticide tolerance regulations fail to protect children. All regulations to date are based on adult tolerances, but safe levels of exposure for adults may be unsafe for children. A study in England and Wales has shown that 50% of all pesticide poisoning incidents in those countries involved children under ten years of age.[21] Use of pesticides in the home is also linked to childhood cancer.[22] Because children's sensitivities to toxicants differ from those of adults, pesticide regulations must be re-evaluated with children in mind.

Although no one can place a precise monetary value on a human life, the economic "costs" of human pesticide poisonings have been estimated. For our assessment, we use the conservative estimate of $2.2 million per human life—the average value that the surviving spouse of a slain New York City police officer receives.[23] Based on the available data, estimates are that human pesticide poisonings and related illnesses in the United States cost about $933 million each year.

Pesticide use, though, provides a substantial net agricultural return of $12 billion per year. Are the public health risks associated with pesticide use a great enough concern to warrant a reduction in pesticide use? A specific example better illustrates this ethical dilemma: assuming that pesticide-induced cancers number 10,000 cases per year, and given the $12 billion per year return from pesticides, each case of cancer is "worth" $1.2 million in pest control. In other words, for every $1.2 million in pesticide benefits, one person falls victim to cancer. Very few of these victims have accepted this health risk voluntarily; the majority of people exposed to the negative health effects of pesticides are not even adequately aware of the risks they face.[24] We as a society must ask, and instruct our policymakers with the answer: Is this ratio of pesticide related cancers to economic benefits an acceptable ratio of human health to economic profit? We have hard scientific data that some—maybe

---

[20] Repetto and Baliga 1996.
[21] Casey and Vale 1994.
[22] Cordier et al. 1994.
[23] Nash 1994.
[24] Lehman 1993.

most—pesticides used today carry with them significant human health risks. Is it morally permissible, then, to expose *any* number of human beings to potential harm, especially if the victims don't voluntarily accept those risks? Or would regulating pesticide use more stringently impinge on the individual rights and food security of the world's citizens? How do we decide what the ethical choice is?

Domestic animal poisonings and contaminated animal products

Several thousand domestic animals are poisoned by pesticides each year, mainly dogs and cats.[25] This is not surprising because they usually wander freely about the home and farm, and therefore have greater opportunity than other domesticated animals to come into contact with pesticides. Estimates indicate that about 20% of the total monetary value of animal production, or about $4.2 billion, is lost to all animal illnesses, including pesticide poisonings.[26] Specifically, about 0.5% of animal illnesses and 0.04% of all animal deaths reported to a veterinary diagnostic laboratory were due to pesticide toxicosis, so about $22 and $9.5 million are lost to pesticide poisonings and pesticide-related deaths, respectively.

This estimate is considered low, though, because it is based only on poisonings reported to veterinarians. Many animal pesticide poisonings that occur in the home and on farms go undiagnosed and are attributed to factors other than pesticide poisonings. In addition, when a farm animal poisoning occurs and little can be done for an animal, the farmer usually waits for the animal to recover or destroys it.[27] Such cases of pesticide poisonings are usually unreported.

Pesticide contamination of animal products—milk, meat and eggs—can lead to even more significant monetary losses. Pesticides applied to feed crops, farm buildings, or directly to the animal for pest control can build up in animal products and render them unsafe for human consumption. Animal products sold for human use are inspected for contamination by the National Residue Program (NRP). However, of the more than 600 pesticides now in use, the NRP tests only for the 41 pesticides that the FDA and EPA have determined to be a public health concern.[28] Animal products determined to contain pesticide residues must be disposed of, sometimes at a tremendous financial loss.

When the costs attributable to domestic animal poisonings and to contaminated meat, milk and eggs are combined, the economic value of all livestock products in the United States lost due to pesticide contamination is estimated to be at least $31.5 million annually.

---

[25] Murphy 1994.

[26] Gaafer et al. 1985.

[27] G. Maylin, Cornell University, PC, 1977.

[28] D. Beerman, Cornell University, PC, 1991.

Similarly, other nations lose significant numbers of livestock and large amounts of animal products each year to pesticide-induced illness or death. Exact data concerning worldwide livestock losses do not exist; the available information comes only from reports of the incidence of mass destruction of livestock.

Countries exporting meat to the United States, though, can experience tremendous economic losses if the meat is found to be contaminated with pesticides. In a 15-year period, the beef industries in Guatemala, Honduras, and Nicaragua lost more than $1.7 million due to pesticide contamination of exported meat.[29] In these countries, meat that is too contaminated for export is sold in local markets. Obviously, such policies contribute to international public health problems and cannot continue.

With an exponentially growing human population, the amount and quality of available food is a significant concern. At present, approximately 3 billion people worldwide are considered malnourished, about half of the total world population.[30] This figure seems staggering, but the number of hungry humans will most likely continue to grow if population and consumption continue to increase at their present rates.[31]

Destruction of beneficial natural predators and parasites
In both the natural and agricultural ecosystems, many species of predators and parasites assist in controlling herbivorous populations. These naturally beneficial species can help ecosystems remain foliated and "green," a quality vital to maintaining environmental health and clean air. Beneficial parasites and natural predators help to keep herbivore populations at low levels and limit plant losses. Natural enemies play a major role in keeping the populations of many insect and mite pests under control.[32]

Like pest species, though, these beneficial natural enemies are adversely affected by pesticides.[33] For example, the following pests have reached outbreak levels in cotton crops following the destruction of natural enemies by pesticides: bollworm, tobacco budworm, cotton aphid, spider mites, and cotton loopers.[34] Significant pest outbreaks have occurred in other crops.[35]

Even low doses of pesticides that don't destroy beneficial insects can negatively impact these natural enemy species. Parasitic and

---

[29] ICAITI 1977.
[30] WHO 1996.
[31] Pimentel et al. 1998.
[32] Huffaker 1977; Pimentel 1988.
[33] Adkisson 1977; Ferro 1987; Jackson & Lam 1989; Croft 1990; Wills et al. 1990.
[34] Adkisson 1977; OTA 1979; Murray 1994.
[35] Huffaker 1977; OTA 1979; Croft 1990; Pimentel et al. 1980; Murray 1994.

predaceous insects often have complex search and attack behaviors; sublethal insecticide dosages or residues may alter this behavior and disrupt this very effective biological control method.[36]

When outbreaks of secondary pests occur because their natural enemies have been destroyed by pesticides, additional—and sometimes more expensive and toxic—pesticide treatments are often required to sustain crop yields, raising overall costs and exacerbating pesticide-related problems. In fact, it is estimated that the destruction of natural enemies by pesticides, the subsequent crop losses, and additional pesticide application cost the United States $520 million every year.[37]

Natural enemies are adversely affected by pesticides worldwide. A striking example is found in the drastic increase in insecticide for rice production in Indonesia from 1980 to 1985,[38] which caused the destruction of beneficial natural enemies of the brown planthopper. The pest population consequently exploded, and rice yields dropped so much that rice had to be imported *into* Indonesia for the first time in many years. The estimated financial loss in just a two-year period was $1.5 billion.[39]

Following that incident, entomologist Dr. I.N. Oka and his associates, who had previously developed a successful, low-insecticide program for rice pests in Indonesia, were consulted by Indonesian President Suharto's staff to determine what could be done to rectify the situation.[40] Their advice was to reduce insecticide use substantially and return to a sound "treat when necessary" program that protected the natural enemies. Following Dr. Oka's advice, President Suharto mandated in 1986 that 57 of 64 pesticides would be withdrawn from use on rice and pest management practices improved. Pesticide subsidies were also eliminated. Subsequently, rice yields increased to levels well above those recorded during the period of heavy pesticide use.[41]

We estimate that while pesticides provide approximately 10% of pest control, natural enemies provide about five times that amount, or half of all pest control.[42] Many cultural controls such as crop rotations, soil and water management, fertilizer management, planting time, crop-plant density, trap crops, and polyculture provide additional pest control. Together, these non-chemical controls can be used effectively to reduce U.S. pesticide use by as much as one-half, without any reduction in crop yields,[43] and can significantly reduce the negative effects of pesticide

---

[36] L.E. Ehler, University of California, PC, 1991.
[37] Pimentel et al. 1993.
[38] Oka 1991.
[39] FAO 1988.
[40] I.N. Oka, Bogor Food Research Institute, Indonesia, PC, 1990.
[41] FAO 1998.
[42] Pimentel et al. 1993.
[43] Pimentel et al. 1991.

use. It would be in the best interest of consumers and producers alike to take steps to protect natural enemies, and it appears they can be protected without a substantial reduction in the global food supply.

Pesticide resistance in pests

The extensive use of pesticides can result in the development of pesticide resistance in insect pests, plant pathogens, and weeds. In a report by the United Nations Environment Programme, pesticide resistance was ranked as one of the top four environmental problems in the world.[44] About 504 insect and mite species,[45] a total of nearly 150 plant pathogen species,[46] and about 273 weed species are now resistant to pesticides.[47] As pesticide use increases, these numbers are sure to climb.

Increased pesticide resistance in pest populations frequently results in the need for several additional applications of commonly used pesticides to maintain expected crop yields. These additional pesticide applications, though, actually compound the problem by increasing environmental selection for resistance. A very small number of insects in a given pest species may be naturally resistant to a particular pesticide, due to a random genetic mutation. The more pesticide applied throughout the environment, the faster pesticide resistance develops, as non-resistant insects are killed off and only resistant insects are left to reproduce. Despite numerous attempts to deal with this problem, pesticide resistance continues to develop at an alarming rate.[48]

One study[49] estimates the costs attributed to pesticide resistance, reporting a yearly loss of $45 to $120 per hectare[50] to pesticide resistance in California cotton. Thus, approximately $348 million of the California crop was lost to resistance. Since $3.6 billion worth of U.S. cotton was harvested in 1984, the loss due to resistance for that year was approximately 10%. Assuming a 10% loss in other major crops that receive heavy pesticide treatments in the United States, crop losses due to pesticide resistance are estimated to be about $1.4 billion per year.

Although the costs of pesticide resistance are high in the United States, costs in many developing countries are significantly greater because pesticides are used not only to control agricultural pests, but are also vital for the control of disease.[51] One of the major impacts of pesticide resistance in tropical countries is associated with malaria control, since the disease is spread largely by mosquitoes. By 1961 the

---

[44] UNEP 1979.

[45] Georghiou 1994.

[46] Eckert 1988.

[47] LeBaron & McFarland 1990.

[48] Murray 1994.

[49] Carrasco-Tauber 1989.

[50] A hectare, abbreviated ha, is about 2.4 acres.

[51] Murray 1994.

annual incidence of malaria in India, after early pesticide use, had declined to only 41,000 cases. However, because mosquitoes developed resistance to pesticides and malarial parasites developed resistance to drugs, the incidence of malaria in India now has exploded to about 59,000,000 cases per year.[52] Similar problems are occurring throughout Asia, Africa, and South America; the annual global incidence of malaria is estimated to be 100 to 120 million cases with 1 to 2 million deaths.[53] Increased pesticide resistance in malaria-carrying mosquitoes is obviously a significant issue.

Pesticide resistance raises an ethical dilemma: we need to use pesticides to control pests that carry disease, but overuse of pesticides results in pesticide resistance and can actually cause the spread of disease. In addition, straightforward economic losses—in the form of crop losses and increased spraying expenditures—to resistant pests are substantial. However, not using pesticides at all would also lead to the spread of disease and huge economic losses. Crops would be destroyed by pests, the world's food supply would be greatly reduced, and pest-borne diseases would surely increase.

Honey bee and wild bee poisonings and reduced pollination

Honey and wild bees are vital for the pollination of fruits, vegetables and other crops. It has been estimated that production of approximately one-third of all human food is dependent on bee pollination.[54] The direct and indirect benefits of bees to agricultural production range from $10 to $33 billion each year in the United States.[55] Because most agricultural insecticides are toxic to bees,[56] pesticides have a major impact on both honey bee and wild bee populations.

Approximately 20% of all honey bee colonies are adversely affected by pesticides, and the yearly estimated loss from partial bee kills, reduced honey production, and the cost of moving colonies, totals about $25.3 million. Also, as a result of heavy pesticide use on certain crops, beekeepers are excluded from 4 to 6 million ha of otherwise suitable apiary locations, and the yearly loss in potential honey production in these regions is about $27 million.[57]

In addition to these direct losses caused by damage to bees and honey production, many crops are lost because of the lack of effective pollinators. Estimates of annual agricultural losses due to the reduction in pollination from pesticides range as high as $4 billion per year.[58] For

---

[52] Reuben 1989.
[53] WHO 1992; NAS 1993.
[54] Williamson 1995.
[55] Robinson et al. 1989; E.L. Atkins, University of California, PC, 1990.
[56] Johansen 1977; MacKenzie & Winston 1989.
[57] D. Mayer, Washington State University, PC, 1990.
[58] J. Lockwood, University of Wyoming, PC, 1990.

most agricultural crops, both yield and quality are enhanced by effective pollination. For several cotton varieties, effective pollination by bees resulted in yield increases from 20% to 30%.[59] Assuming that a conservative 10% increase in cotton yield would result from more efficient pollination, and subtracting charges for bee rental, the net annual gain for cotton alone could be as high as $400 million. However, using bees to enhance cotton pollination is impossible at present because of the intensive use of insecticides on cotton.[60]

Adequate pollination provides similar benefits to food crops like fruits and vegetables. For example, with adequate pollination, melon yields were increased 10% and quality was raised 25%, as measured by the dollar value of the crop.[61]

Based on the analysis of honey bee and pollinator losses caused by pesticides, pollination losses attributed to pesticides are about 10% of pollinated crops, at a yearly cost of about $200 million. The combined annual costs of reduced pollination and direct loss of honey bees due to pesticides can be estimated at about $320 million.

With the world's human population increasing, we cannot afford to lose 10% of pollinated crops to pesticide effects. Regulations to protect natural pollinators, as well as beneficial natural enemies, may cause a reduction in individual autonomy for farmers and pesticide manufacturers, but this loss of a fraction of personal freedom for a few may mean the preservation of the ability to feed the people of the world.

Crop and crop product losses

Pesticides are applied to protect crops from pests in order to increase harvests, but sometimes the crops themselves are damaged by pesticide treatment.[63-66] Damage occurs when: (1) the recommended dosages suppress crop growth, development and yield; (2) pesticides drift from the targeted crop to damage adjacent crops; (3) residual herbicides either prevent chemical-sensitive crops from being planted in rotation or inhibit the growth of crops that are planted; and/or (4) excessive pesticide residues accumulate on crops, necessitating the destruction of the harvest. Crop losses translate into financial losses for growers, distributors, wholesalers, transporters, retailers, food processors, and others. Potential profits as well as investments are lost. The costs of crop losses increase even further when the related costs of investigations, regulation, insurance, and litigation are added. Ultimately, the consumer pays for these losses in higher prices at the marketplace.

---

[59] McGregor et al. 1955; Tanda 1984.

[60] McGregor 1976.

[61] E.L. Atkins, University of California, PC, 1990.

Heavy and even normal dosages of insecticides have been reported to suppress growth and yield in crops.[62,63] Normal use of some types of herbicides can cause increased susceptibility to insects and diseases,[64] and when weather or soil conditions are inappropriate for pesticide application, herbicide treatments can seriously weaken crops and cause yield reductions ranging from 2% to 50%.[65]

Crops can also be destroyed when pesticides drift from the target crops to nontarget crops, sometimes as far as several miles downwind.[66] Although drift occurs with almost all methods of pesticide application, including both ground and aerial equipment, the problem is greatest when pesticides are applied by aircraft. With aircraft applications, as much as 50% to 75% of pesticides applied miss the target area.[67] In contrast, 10% to 35% of the pesticide applied with ground application equipment misses the target area.[68] Crop injury and subsequent loss due to drift is particularly common in areas planted with diverse crops. For example, in southwest Texas in 1983 and 1984, nearly $20 million of cotton was destroyed from drifting 2,4-D herbicide when adjacent wheat fields were aerially sprayed.[69] Crop losses due to pesticide drift are financially damaging to crop producers and ultimately to the consumers. In addition, drifting pesticides can negatively impact large areas of the environment and natural ecosystems.

An average of 0.1% loss to direct pesticide use in the annual U.S. production of corn, soybeans, cotton and wheat—crops that together account for about 90% of the herbicides and insecticides used in U.S. agriculture—was valued at $40.9 million in 1993.[70] Assuming that only one third of the incidents involving crop losses due to pesticides are actually reported to authorities (since many incidents are settled privately), the total value of all crops lost because of pesticides could be as high as three times this amount, or $123 million annually.

Additional economic and social costs are incurred when food crops are disposed of because they exceed the EPA regulatory tolerances for pesticide residue levels. Assuming that all the crops and crop products that exceed the EPA regulatory tolerances (reported to be at least 1%) were disposed of as required by law, then about $550 million

---

[62] ICAITI 1977; Trumbel et al. 1988.

[63] J. Neal, Chemical Pesticides Program, Cornell University, PC, 1990.

[64] Oka & Pimentel 1976; Altman 1985; Rovira & McDonald 1986.

[65] Von Rumker & Horay 1974; Elliot et al. 1975.

[66] Barnes et al. 1987.

[67] ICAITI 1977; Ware 1983.

[68] Ware et al. 1975; Hall 1991.

[69] Hanner 1984.

[70] USDA 1994.

in crops would be destroyed annually because of excessive pesticide contamination.[71]

When crop seizures, insurance and investigation costs are added to the costs of direct crop losses due to the use of pesticides in commercial crop production, the total monetary loss is estimated to be about $959 million annually in the United States alone. These huge financial losses are equally substantial: crop losses due to pesticide use result in a decreased food supply and significant health risks when people eat food products containing undetected pesticides. However, cessation of pesticide use would also lead to economic and humanitarian misfortune. And increased testing for pesticide residues is expensive as well—the cost of state investigations alone is estimated at $10 million annually.[72]

Groundwater and surface water contamination

Most soluble pesticides applied to crops eventually drift and/or erode into ground and surface waters, and some pesticides have even been detected in rain and fog. It has been estimated that the total runoff of all pesticides from non-irrigated farmland is 1%, runoff from irrigated farmland is 4%, and the runoff from airplane-applied pesticides is 33%.[73]

Pesticide contamination of surface water—lakes, rivers and streams—is a serious concern because surface water is extensively used for drinking and recreation. We should be particularly alarmed that conventional drinking water treatment is not designed for, *and therefore does not remove*, pesticides. One study showed that after conventional treatment, 90% of drinking water samples contained at least one pesticide, while 58% of the samples contained at least four different pesticides.[74] After studying the drinking water in 29 U.S. cities over a four-month period, Cohen et al.[75] observed that the herbicide atrazine was present in the tap water of all but one of the cities (97%) and cyanazine was present in all but four (86%). In addition, federal health levels for atrazine and cyanazine were exceeded in 17% and 35% of all samples taken, respectively. At present, water treatment plants do not regularly monitor for pesticide contamination. Estimates for the daily monitoring cost of the triazine herbicides is $1500 per city per year, or ten cents per person,[76] totaling approximately $27 million per year.[77]

In addition, estimates suggest that nearly one-half of the groundwater and well water in the United States is or has the potential to

---

[71] Calculated based on data from FDA (1990) and USDA (1989).
[72] S.F. Baril, State of Montana, Dept. of Agriculture, PC, 1990.
[73] Anayeva et al. 1992.
[74] Kelley 1986.
[75] Cohen et al. 1995.
[76] Id.
[77] Natarajan & Rajagopal 1994.

be contaminated.[78] EPA reported that 10.4% of community wells and 4.2% of rural domestic wells have detectable levels of at least one of the 127 pesticides tested in a national survey.[79] Specifically, groundwater contaminated with pesticides is a huge problem for two major reasons: first, about one-half of the population obtains its water from wells,[80] and second, once groundwater is contaminated, the pesticide residues remain for a long period of time.[81] Not only are there just a few microorganisms that have the potential to degrade pesticides,[82] but the groundwater is replenished at an average rate of less than 1% per year. It would cost an estimated $1.3 billion annually, though, if well and groundwater were monitored for pesticide residues in the United States.[83]

Monitoring groundwater for pesticide contamination is only a portion of the total cost of groundwater contamination; cleaning up the polluted water also exacts a high price. It is estimated that if all pesticide-contaminated groundwater were cleared of pesticides before human consumption, the cost would be about $500 million (based on the costs of cleaning water.[84] The cleanup process first requires a water survey to target the contaminated water for cleanup. Thus, by adding the monitoring costs to the cleaning costs, the total cost regarding pesticide-polluted groundwater is estimated to be about $1.8 billion annually.

$1.8 billion, though, might be a small price to pay compared with the cost of continued pollution of our water resources, Some of the pesticides found in drinking water—atrazine, for instance—are known mutagens or carcinogens.[85] These compounds should be tested for and removed despite the cost. The world's water supply is a limited resource, and we cannot survive as individuals or as a species without clean drinking water. While the financial costs of testing for and removing pesticides from our water supply may seem imposing at present, the economic and social costs will only increase if the world's supply of clean water continues to decrease.

Fishery losses

As described in the previous section, pesticides are frequently washed or blown into aquatic ecosystems by water runoff and soil erosion. In addition to contaminating the water itself, pesticides in aquatic ecosystems can also affect the organisms living there. Some

---

[78] NRDC 1993.
[79] EPA 1990a.
[80] Beitz et al. 1994.
[81] Gustafson 1993.
[82] Pye & Kelley 1984.
[83] Nielsen & Lee 1987.
[84] Van der Leeden et al. 1990.
[85] Culliney at al. 1993.

soluble pesticides are easily leached into streams and lakes[86] and are readily taken up by aquatic organisms.[87] A nationwide survey of U.S. fish showed pesticide residues in almost all of the 119 fish species examined.[88]

Once in aquatic systems, pesticides cause fishery losses in several ways. High pesticide concentrations in water can directly kill fish; low-level doses may kill highly susceptible fish or weaken other fish; the elimination of essential fish foods like insects and other invertebrates can result in the loss of entire fish species, as can the reduction of dissolved oxygen levels in the water due to the decomposition of aquatic plants killed by herbicides. In addition, because government safety restrictions ban the catching or sale of pesticide residue-contaminated fish, such unmarketable fish are considered an economic loss.[89]

Each year large numbers of fish are killed by pesticides. Based on EPA data,[90] we calculate that from 1977 to 1987 the number of fish lost to all factors was 141 million fish per year; the number of fish killed by pesticides is estimated at 6 to 14 million yearly, certainly a conservative estimate.[91]

Using the estimated value of $4 per fish, the value of the conservative estimate of 6 to 14 million fish killed each year ranges from $24 to $56 million. This calculation takes into account only direct impacts. The costs of direct impacts are more difficult to define but can be substantial as well. For instance, the revenue generated in Massachusetts by marine recreational fishing has been estimated at $637 million annually.[92] Thus, when all indirect impacts are taken into account, the actual financial loss is probably several times the estimated direct cost of $24 to $56 million, or well over $100 million.

Fish kills due to pesticide contamination of water raise ethical concerns similar to those of crop and pollinator losses: our already inadequate global food supply is further reduced when fish are lost, and people who make their living catching, raising or selling fish are hurt economically. As the supply of fish becomes more limited, their cost goes up and consumers suffer as well. In addition, the loss of large numbers of fish, sometimes the loss of entire species, raises a new environmental concern: the conservation of biodiversity. The health of the world's ecosystems depends on a delicate balance of species, of predators and prey, a numbers game between individuals and available

---

[86] Nielsen & Lee 1987.
[87] Buttner et al. 1995; Chevreuil et al. 1995.
[88] Kuehl & Butterworth 1949.
[89] EPA 1990b.
[90] Id.
[91] EPA 1990b.
[92] Storey & Allen 1993.

resources. When humans disrupt the delicate balance of even a single population, species or ecosystem, the health of the world's environment is impacted. There are thousands of diverse species in the world, each with a unique niche. We cannot know the precise long-term impacts the destruction of even one vital species may have on the global environment; disregard for nature's system of checks and balances may seriously endanger our own livelihood and well-being. Therefore, steps towards the conservation of biodiversity may bring important societal benefits.

Wild birds and mammals

Like fish, wild birds and mammals are also damaged by pesticides; these animals, like the canary in the coal mine, make excellent "indicator species"—species whose numbers, if falling, can indicate an environmental problem. Deleterious effects on wildlife include death from direct exposure to pesticides or secondary poisonings from consuming contaminated prey; reduced survival, growth and reproductive rates from exposure to sublethal dosages; and habitat reduction through elimination of food sources and refuges.[93] In the United States, approximately 3 kg of pesticide per hectare are applied on about 160 million ha of land per year.[94] With such a large portion of the land area treated with heavy dosages of pesticide, the impact on wildlife is expected to be significant. The full extent of bird and mammal destruction is difficult to determine exactly, because animals are often secretive, camouflaged, highly mobile, and live in dense grass, shrubs and trees.

Nevertheless, many bird casualties caused by pesticides have been reported. Some highly toxic or heavily applied pesticides can kill birds and other animals directly. Carbofuran applied to vegetable crops killed more than 5000 ducks and geese in 5 incidents, while the same chemical applied to vegetable crops killed 1400 ducks in a single incident.[95] Carbofuran is estimated to kill 1 to 2 million birds each year.[96] Another pesticide, diazinon, applied on just three golf courses, killed 700 of the wintering population of 2500 Atlantic Brant geese.[97]

In addition to directly killing birds, the indirect effects of pesticide applications—such as the elimination of food supply or habitats, and/or reproductive effect—can adversely affect bird and wildlife species. Several studies report that the use of herbicides in crop production results in the total elimination of weeds that harbor some

---

[93] Pimentel et al. 1993.
[94] Pimentel et al. 1991.
[95] Flickinger et al. 1980; 1991.
[96] EPA 1989.
[97] Stone & Gradoni 1985.

insects, which subsequently causes a reduction in the insect population as well.[98] Loss of certain insects has led to a 77% reduction in the grey partridge population in the United Kingdom, and a comparable decrease in the numbers of common pheasants in the United States. These substantial reductions in bird populations occurred because the partridge and pheasant chicks depend on insects to supply them with needed protein for their development and survival.[99] Pesticides and herbicides used on nearby crops reduced the habitat of the local insects and drastically limited the birds' food supply.

Pesticides also adversely affect the reproductive potential of many birds and mammals. Exposure of birds, especially predatory birds, to chlorinated insecticides has caused reproductive failure, sometimes attributed to eggshell thinning.[100] Most of the affected populations recovered after the ban of DDT in the United States.[101] However, DDT and its metabolite DDE remain a concern even for many North American birds, because DDT is still used in some South American countries where numerous bird species winter.[102] In another well-known example, the decline of the alligator population in Florida's Lake Apopka was linked to a large spill of DDT. Not only were there problems with eggs in every nest examined, but researchers observed that the male alligators had abnormally low testosterone levels and that 25% of them had shrunken reproductive organs that would never allow them to reproduce.[103]

These findings have, appropriately, raised concern about how humans are affected by these chemicals, especially since the average human male sperm count has decreased worldwide by more than 40% since 1938.[104] The possible threat certainly warrants new, thorough research efforts.

Although gross economic values for wildlife are not available, the money that humans spend on wildlife-related activities are one measure of the monetary value of wild birds and mammals. Nonconsumptive users of wildlife spent an estimated $18 billion on their activities per year.[105] Each year, U.S. bird watchers spend an estimated $600 million on their sport and an additional $500 million on birdseed, a total expenditure of $1.1 billion.[106] The money spent by bird hunters to

---

[98] R. Beiswenger, University of Wyoming, PC, 1990.
[99] Potts 1986; R. Beiswenger, University of Wyoming, PC, 1990.
[100] Pimentel et al. 1993.
[101] Bednarz et al. 1990.
[102] Stickel et al. 1984.
[103] Begley & Glick 1994.
[104] Carlsen et al. 1992.
[105] USDI 1991.
[106] USFWS 1988.

harvest 5 million game birds was reported to be $1.1 billion, or approximately $216 per bird.[107]

If we assume that the damages pesticides inflict on birds occur primarily on the 160 million hectares of cropland that receive most of the pesticides applied, and if the bird population is estimated to be 4.2 birds per hectare of cropland,[108] then 672 million birds are directly exposed to pesticides. If it is conservatively estimated that only 10% of the bird population is killed, then the total number killed each year is 67 million birds. This is considered a conservative estimate because secondary losses to pesticide reductions in invertebrate prey poisonings are not included in the assessment. Estimates for the value of all types of birds ranged from $0.40 to more than $800 per bird.[109] Assuming the average value of a bird is $30, then an estimated $2 billion in birds are destroyed by pesticides each year.

Like fish kills, bird and wildlife losses caused by pesticides decrease the earth's essential biodiversity. Even if humans do not value wildlife species in and of themselves, we should be aware that by destroying their habitats and reducing their numbers, we are harming the entire planet, *our own habitat*. Protecting our habitat certainly is essential for the whole human population, which cannot survive if our habitat is destroyed. The reduction of wildlife by pesticide use, then, is ethically irresponsible.

Microorganisms and invertebrates

Pesticides easily find their way into soil, where they may be toxic to the arthropods, earthworms, fungi, bacteria, and protozoa found there. These small organisms are vital to ecosystem survival because they dominate both the structure and function of natural systems. Fungi, bacteria, and arthropods make up 95% of all species in the earth's soil and 98% of the biomass (excluding vascular plants). These microorganisms and invertebrates are essential to the proper functioning of the ecosystem because they break down organic matter and enable the vital chemical elements to be recycled.[110] Equally important is their ability to "fix" nitrogen and make it available for plants to use and grow and produce oxygen.[111]

Unfortunately, our understanding of the effects of pesticides on soil organisms is minimal. Studies have shown a wide range of pesticide effects, ranging from drastically lowered survival rates to, in some

---

[107] Id.

[108] Blew 1990.

[109] Pimentel et al. 1993.

[110] Atlas & Bartha 1987.

[111] Brock & Madigan 1988.

instances, stimulated growth.[112] These varied reactions to pesticides are not surprising since invertebrates and microorganisms are a large and diverse group of species; each individual species reacts differently to changes and circumstances in its environment. However, in general pesticides—insecticides, fungicides and herbicides—reduce both the diversity of these invertebrate species in the soil, as well as the total biomass of these important biota.[113]

The leaves of apple trees accumulate on the surface of the soil where earthworms were killed by pesticides. Apple scab, a disease carried over from season to season on fallen leaves, is commonly treated with fungicides. Some of these fungicides are toxic to the earthworms, preventing them from removing and recycling the surface leaves[114,115] Earthworms are crucial for the recycling and return of nutrients to the soil; if earthworms are poisoned by pesticides and are unable to perform this duty, many other organisms that depend on these nutrients are harmed as well.

Although these invertebrates and microorganisms are essential to the vital structure and function of all ecosystems, it is impossible to place a dollar value on the damage caused by pesticides to this large group of organisms. To date no relevant quantitative data on the value of microorganism destruction has been collected. The loss of even a small number of microorganism or invertebrate species due to pesticides could be disastrous for the planet if vital steps in the recycling of wastes and nutrients are affected. Limiting pesticide use to preserve microorganisms and invertebrates may serve to benefit the health of the entire planet.

Conclusion and ethical analysis

Although pesticides currently provide about $16 billion per year in saved crops in the U.S., data suggest that the environmental and social costs of pesticides to the nation total more than $8 billion. From a strictly short-term cost/benefit approach, it appears that pesticide use is beneficial. However, the ongoing negative environmental and health impacts of pesticide use may ultimately result in much greater economic and social costs. Weighing the environmental and health impacts of pesticide use against the economic benefits of pest control requires a serious look at the ethical issues surrounding pesticide use.

We will focus on three common ethical arguments or approaches that relate most clearly to the main areas of concern related to pesticide use—public health, environmental, and economic effects of pesticides.[116]

---

[112] Edwards 1989; Jones et al. 1991.
[113] Pimentel et al. 1997.
[114] Stringer & Lyons 1974.
[115] Edwards & Lofty 1977.
[116] Lehman 1993; Perkins & Holochuck 1993.

There are essentially three possible courses of action we can take with regard to these areas of concern: (1) dismiss these concerns and continue the widespread and increasing use of pesticides; (2) stop using synthetic pesticides altogether to protect the public health and the environment; or (3) weigh the social, environmental and economic costs against the many benefits of pesticide use, and develop a strategy of pest management designed to maximize the benefits and minimize the costs. An examination of the ethical issues related to pesticide use will help evaluate these three options.

As the earlier discussions in this paper indicate, pesticides in general pose significant health risks for people exposed to them, especially children, and even unborn infants. Pesticides have been shown to affect reproductive cells and processes in other animals;[117] if reproductive processes are affected in humans to the same extent, then the pesticides used today have the potential to impact future generations of human beings decades from now. And most people exposed to the health effects of pesticides do not voluntarily choose to accept that risk; rather, they are unwittingly exposed to pesticides drifting from target areas or pesticide residue in contaminated foods. While the economic costs of these health-related consequences are too often not considered to be as significant as the economic benefits of pesticide use, the moral costs are substantial: should the potential health risks today and into the future preclude the use of pesticides despite the obvious immediate economic benefits for a few?

Resolving this question from a purely consequentialist viewpoint is difficult because the data related to health risks are not complete. One could argue that fewer people suffer illness and death as a result of pesticide exposure than the substantial number of people worldwide who benefit from the food produced using pesticides. In addition, eliminating pesticide use might result in enormous economic losses for hundreds of thousands of people, perhaps harming more people than are made ill by pesticides. So initially, we might conclude that more people benefit from pesticides than are harmed by them; therefore, continued pesticide use would be the moral choice, despite the potential health risk, because pesticides provide the greatest good for the greatest number—for the time being. With pesticide resistance and the destruction of natural enemies increasing, though, the amount of pesticides used worldwide— and the number of people exposed to their ill effects—may increase as well. It is conceivable that at some point in the future, more people may suffer from pesticide-related illness than will benefit from the food that pesticides help produce. In relation to public health concerns, then, a long-term consequentialist argument suggests that the current trend in

---

[117] Begley & Glick 1994.

pesticide usage will not serve to maximize utility and benefit the greatest number of people.

Both the Kantian categorical imperative and a libertarian ethical theory support this conclusion. In Kantian terms, the smaller number of people who suffer ill health effects due to pesticide exposure are being treated as a means to economic benefit rather than as an end in themselves, a means to producing a greater food supply and preventing economic losses to farmers and pesticide manufacturers. The Kantian argument would posit that no amount of health risk is acceptable because those who fall victim to those risks are not expendable for the greater good, but are valuable in and of themselves.

In addition, an ethical theory focused on the untouchable autonomy of the individual would claim that inadvertent exposure to pesticides compromises the individual autonomy of the victims—they did not choose to be exposed to pesticides, nor were they fully informed of the potential risks involved. Thus, these various ethical approaches lead to the consistent single conclusion that the ethical choice for pest management is one that considers the protection of public health as a top priority.

Pesticides have the potential to disrupt entire ecosystems. When they destroy beneficial natural predators of common and secondary pests, or a vital pollinator species, pesticides throw off the delicate balance and interrelatedness of nature, and often lead to the use of even greater amounts of still more toxic pesticides. Heavy use of pesticides can result in a multitude of consequences, from pesticide resistance to surface water contamination and wildlife poisonings. All these effects, directly or indirectly, can serve to decrease vital biodiversity and further disturb natural ecosystems. As residents of this planet, we cannot help but be affected by these environmental effects; to thrive, human beings require clean water, clean air, and an adequate food supply, just as all other living organisms do.

As the extensive data discussed in this paper suggest, pesticide use is not without significant environmental consequences, consequences that directly impact our lives. It is in the best interests of the entire human race to conserve resources and protect the environment, yet we are knowingly polluting and endangering the environment with even our current levels of pesticide use. The vast majority of the environmental effects of pesticide use are negative, harming the environment; since we need a flourishing environment to survive as individuals and as a species, it seems only reasonable that the greatest good for the greatest number will come from protecting and preserving the environment—our habitat.

Taking steps to protect the environment would most likely result in stricter pesticide regulations and more federal involvement in pesticide use, and therefore some limitations on personal freedom for a few. One would probably not be free to apply highly toxic pesticides anywhere and in any way, for one's own gain, as egocentric ethics would suggest is one's right. However, given our greater understanding of the negative health and environmental effects of pesticides, one might well equate the wanton use of pesticides with directly and indirectly harming other people. The use of pesticides might then be construed as treating all the people on the planet as the means to an end of personal gain, an unethical action regardless of consequences or individual autonomy. A small limitation on personal freedom seems a reasonable price to pay when the social and ethical costs of continued environmental destruction are considered.

Finally, we must address the economic issues related to pesticide use. An investment of about $6.5 billion in pesticide control saves approximately $26 billion in U.S. crops, based on direct costs and benefits.[118] However, the indirect costs of pesticide use to the environment and public health need to be balanced against these benefits. A major cost associated with all pesticide use is the cost of carrying out state and federal regulatory actions, as well as the pesticide monitoring programs needed to control pesticide pollution. Specifically, these funds are spent to reduce the hazards of pesticides and to protect the integrity of the environment and public health.

In addition, at least $1 million is spent each year by the federal and state governments to train and register pesticide applicators.[119] Also, more than $40 million is spent each year by the EPA just for registering and re-registering pesticides.[120] Based on these known expenditures, the federal and state governments are estimated to spend approximately $200 million per year for pesticide pollution control (Table 2).[121]

Although enormous amounts of government money are being spent to reduce pesticide pollution, many costs of pesticides are not taken into account. Based on the available data, a conservative estimate of the environmental and social costs of pesticide use is approximately $8.3 billion each year (Table 2). Users of pesticides in agriculture pay only about $3.2 billion of this cost. Society eventually pays this $3.2 billion plus the remaining $5.1 billion in environmental and public health costs (Table 2). And if pesticide use continues to increase, these costs are sure to increase as well.

Ethical considerations of the social, environmental and economic impacts of the current usage of pesticides suggest that we cannot continue our present levels and methods of pesticide use, despite the overall economic benefits at present. However, the abrupt and complete cessation of synthetic herbicide use is not without substantial costs of its

---

[118] Pimentel et al. 1991; USBC 1994.
[119] D. Rutz, Cornell University, PC, 1991.
[120] GAO 1986.
[121] USBC 1994.

own. Lehman[122] describes the huge crop losses which would result from the termination of pesticide use. Surely, the greatest good for the greatest number is not served by a sudden, complete elimination of pesticides from global agriculture. So what should we do?

Strong arguments can be made to support pesticide use because it has quantifiable social and economic benefits. However, these benefits should not conceal the public health and environmental problems associated with pesticide use. Our goal should be to maximize the benefits while at the same time minimizing the health, environmental and social costs of pesticides. Clearly, it is essential that the environmental and social costs and benefits of pesticide use be considered when future pest control programs are being developed and evaluated. Such costs and benefits should be given ethical and moral scrutiny before policies are implemented, so that sound, sustainable pest management practices are available to benefit farmers, society and the environment. We agree with Lehman's recommendation that a gradual reduction in the global use of synthetic pesticides is the most reasonable, beneficial and ethical way to reach this goal.

Judicious, responsible use of pesticides could reduce the environmental and social costs, while benefiting farmers economically in the short run and supporting the sustainability of agriculture in the long run. Pesticide manufacturers would suffer some economic losses, but no widespread or social damage would be done, as studies have shown that nonchemical controls, such as natural enemies and crop rotations, can reduce U.S. pesticide use by up to one-half without any reduction in crop yields.[123] Pesticides are and will continue to be a valuable pest control tool. Meanwhile, with more accurate and realistic cost/benefit analyses, we will be able to work to minimize risks and increase the use of non-chemical pest controls to maximize the benefits of pest control strategies for all of society.

---

[122] Lehman 1993.
[123] Pimentel et al. 1991.

Table 1.
Estimated economic costs of human pesticide poisonings and other pesticide-related illnesses in the United States each year.

Human health effects from pesticides[a]	Total costs ($)
Cost of hospitalized poisonings 2380 x 2.84 days @ $1000/day	6,759,000
Cost of outpatient treated poisonings 27,000[b] x $630	17,010,000
Lost work due to poisonings 4680 workers x 4.7 days x $80/day	1,760,000
Pesticide cancers <12,000[c] cancers x $70,700[c]/case	848,400,000
Cost of fatalities 27 accidental fatalities[b] x $2.2 million	59,400,000
TOTAL	933,329,000

a=Includes hospitalization, foregone earnings, and transportation.
b=J. Blondell, 1991 personal communication , EPA, Washington, DC.
c=See text for details.

Table 2.
Total estimated environmental and social costs from pesticides in the United States.

Costs	Millions of $/year
Public health impacts	933
Domestic animal deaths and contamination	31
Loss of natural enemies	520
Cost of pesticide resistance	1,400
Honey bee and pollination losses	320
Crop losses	959
Surface water monitoring	27
Groundwater contamination	1,800
Fishery losses	56
Bird losses	2,100
Government regulations to prevent damage	200
TOTAL	8,346

References

Adkisson, P.L. 1977. Alternatives to the unilateral use of insecticides for insect pest control in certain field crops. Edited by L.F. Seatz. Symposium on Ecology and Agricultural Production. Knoxville: University of Tennessee, 129-144.

Atlas, R.M. and R. Bartha. 1987. Microbial Biology: Fundamentals and Applications. 2nd ed. Menlo Park, CA: Benjamin Cummings Co.

Barnes, C.J., T.L. Lavy and J.D. Mattice. 1987. Exposure of non-applicator personnel and adjacent areas to aerially applied propanil. Bull. Environ. Contam. and Tox. 39:126-133.

Bednarz, J.C., D. Klem, L.J. Goodrich and S.E. Senner. 1990. Migration counts of raptors at Hawk Mountain, Pennsylvania, as indicators of population trends, 1934-1986. Auk 107:96-109.

Begley, S., and D. Glick. 1994. The estrogen complex. Newsweek, March 21: 76-77.

Beitz, H., H. Schmidt, and F. Herzel. 1994. Occurrence, toxicological and ecotoxicological significance of pesticides in groundwater and surface water. Edited by H. Borner. Pesticides in Ground and Surface Water. Berlin: Springer-Verlag, 1-56.

Benbrook, C.M., E. Groth, J.M. Hoaaloran, M.K. Hansen, and S. Marquardt. 1996. Pest Management at the Crossroads. Yonkers, NY: Consumers Union.

Blew, J.H. 1990. Breeding Bird Census. 92 Conventional Cash Crop Farm. Jour. Field Ornithology. 61 (Suppl.) 1990: 80-81.

Brock, T. and M. Madigan. 1988. Biology of Microorganisms. London: Prentice Hall.

Buttner, J.K., J.C. Makarewickz, and T.W. Lewis. 1995. Concentration of selected priority organic contaminants in fish maintained on formulated diets in Lake Ontario water. The Prog. Fish-Culturist 57: 141-146.

Carlsen, E., A. Giwercman, N. Kielding, N.E. Skakkebaek. 1992. Evidence for decreasing quality of semen during past 50 years. Brit. Med. Jour. 305:609-613.

Carrasco-Tauber, C. 1989. Pesticide Productivity Revisited. Amherst: M.S. Thesis. University of Massachusetts.

Casey, P. and J.A. Vale. 1994. Deaths from pesticide poisoning in England and Wales: 1945-1989. Human & Exp. Toxicol. 13:95-101.

Cohen, B., Wiles, and E. Bondoc. 1995. Weed Killers by the Glass. Washington, DC: Environmental Working Group.

Cordier, S., M.J. Iglesias, C. LeGoaster, M.M. Guyot, L. Mandereau, and D. Heman. 1994. Incidence and risk factors for childhood brain tumors in the Ile de France. Intl. J. of Cancer 59: 776-782.

Croft, B.A. 1990. Arthropod Biological Control Agents and Pesticides. New York: Wiley.

Eckert, J.W. 1988. Historical development of fungicide resistance in plant pathogens. Ed. C.J. Delp. Fungicide Resistance in North America. St. Paul: APS Press, 1-3.

Edwards, C. and J. Lofty. 1977. Biology of Earthworms. London: Chapman and Hall.

Elliot, B.R., J.M. Lumb, T.G. Reeves, and T.E. Telford. 1975. Yield losses in weed-free wheat and barley due to post-emergence pesticides. *Weed Res.* 15-107-111.

EPA. 1989. Carbofuran: A Special Review Technical Support Document. Washington, DC: U.S. Environmental Protection Agency, Office of Pesticide and Toxic Substances.

EPA. 1990a. National Pesticide Survey—Summary. Washington, DC: U.S. Environmental Protection Agency.

EPA. 1990b. Fish Kills Caused by Pollution: 1977-1987. Washington, DC: Draft Report of U.S. Environmental Protection Agency. Office of Water Regulations and Standards.

FAO. 1988. Integrated Pest Management in Rice in Indonesia. Jakarta: Food and Agriculture Organization. United Nations. May.

FDA. 1990. Food and Drug Administration Pesticide Program Residues in Foods—1989. *Jour. Assoc. Off. Anal. Chem.* 73:127A-146A.

Flickinger, E.L., G. Juengr, T.J. Roffe, M.R. Smith, and R.J. Irwin. 1991. Poisoning of Canada geese in Texas by parathion sprayed for control of Russian wheat aphid. *Jour. Wildlife Diseases* 27:265-268.

Flickinger, E.L., K.A. King, W.F. Stout, and M.M. Mohn. 1980. Wildlife hazards from furadan 3G applications to rice in Texas. *Jour. Wildlife Mgt.* 44:190-197.

GAO. 1986. Pesticides: EPA's Formidable Task to Assess and Regulate their Risks. Washington, DC: U.S. General Accounting Office.

Georghiou, G.P. 1994. Principles of insect resistance management. *Phytoprotection* 75:51-59.

Gustafson, D.I. 1993. Pesticides in Drinking Water. New York: Van Nostrand Reinhold.

Hall, F.R. 1991. Pesticide application technology and integrated pest management (IPM). Ed. D. Pimentel. Handbook of Pest Management in Agriculture. Vol. II. Boca Raton: CRC Press, 135-170.

Hanner, D. 1984. Herbicide drift prompts state inquiry. *Dallas Morning News*, July 25.

Huffaker, C.B. 1977. Biological Control. New York: Plenum.

ICAITI. 1977. *An Environmental and Economic Study of the Consequences of Pesticide Use in Central American Cotton Production.* Guatemala City, Guatemala: Final Report, Central American Research Institute for Industry, United Nations Environment Programme.

Johansen, C.A. 1977. Pesticides and pollinators. *Annu. Rev. Entomol.* 22:177-192.

Jones, A.L., D.B. Johnson, and D.L. Suett. 1991. Effects of soil treatments with aldicarb, carburofan and chlorfenvinphos on the size and composition of microbial biomass. Ed. A.

Walker. Pesticides in Soils and Water: Current Perspectives. United Kingdon: British Crop Protection Council, 75-82.

Kelley, R.D. 1986. Pesticides in Iowa's drinking water. Proceedings of a Conference: Pesticides and Groundwater: A Health Concern for the Midwest. Navarre, MN: Freshwater Foundation, Oct. 16-17. 121-122.

Kuehl, D.W. and B. Butterworth. 1994. A national study of chemical residues in fish III: Study results. *Chemosphere* 29:523-535.

Lebaron, H.M. and J. McFarland. 1990. Herbicide Resistance in Weeds and Crops. Ed. M.B. Green, H.M. Lebaron, and W.K. Moberg. Managing Resistance from Fundamental research to Practical Strategies. Washington, DC: Am. Chem. Soc., 336-352.

Lehman, H. 1993. Values, Ethics and the Use of Syntheic Pesticides in Agriculture. Ed. D. Pimentel and H. Lehman. The Pesticide Question: Environment, Economics, and Ethics. New York: Chapman & Hall, 347-379.

Leiss, J.K. and D.A. Savitz. 1995. Home pesticide use and childhood cancer: A case-control study. *Am. J. Public Health* 85:249-252.

Litovitz, T.L., B.F. Schmitz, and K.M. Bailey. 1990. 1989 Annual report of the American Association of Poison Control Centers National Data Collection Centers. *Am. J. Emergency Med.* 8: 394-442.

MacKenzie, K. and M.L. Winston. 1989. Effects of sublethal exposure to diazinon and temporal division of labor in the honeybee. *J. Econ. Entom.* 82:75-82.

McCarthy, S. 1993. Congress takes a look at estrogenic pesticides and breast cancer. *Journal of Pesticide Reform* 13 (4): 25.

McGregor, S.E. 1976. Insect Pollination of Cultivated Crop Plants. Washington, DC: U.S. Dept. of Agr., Agr. Res. Ser., Agricultural Handbook No. 496.

McGregor, S.E., C. Rhyne, S. Worley, and F.E. Todd. 1955. The role of honeybees in cotton pollination. *Agron. Jour.* 47:23-25.

Murphy, M.J. 1994. Toxin exposures in dogs and cats: Pesticides and biotoxins. *J. Am. Vet. Med. Assoc.* 205:414-421.

Murray, D.L. 1994. Cultivating Crisis: The Human Cost of Pesticides in Latin America. Austin: Univ. of Texas Press.

NAS. 1987. Regulating Pesticides in Food. Washington, DC: National Academy of Sciences.

NAS. 1993. Malaria Prevention and Control. Washington, DC: National Academy of Sciences.

Nash, E.P. 1994. What's a life worth? New York: New York Times.

Natarajan, U., and R. Rajagopal. 1994. Economics of screening for pesticides in ground water. *Water Resour.Bull.* 30:579-588.

106

Nielsen, E.G. and L.K. Lee. 1987. The Magnitude and Costs of Groundwater Contamination from Agricultural Chemicals: A National Perspective. Washington, DC: U.S. Dept. of Agr., Econ. Res. Ser., Natural Resour. Econ. Div., ERS Staff Report, AGES870318.

NRDC. 1993. *After Silent Spring*. New York: NRDC Publication.

Oka, I.N. 1991. Success and challenges of the Indonesian national integrated pest management program in the rice-based cropping system. *Crop Protection* 10:163-165.

Oka, I.N. and D. Pimentel. 1976. Herbicide (2,4-D) increases insect and pathogen pests on corn. *Science* 193:239-240.

OTA. 1979. Pest Management Strategies. Washington, DC: Office of Technology Assessment, U.S. Congress. 2 Vol.

Perkins, J.H. and N.C. Holochuck. 1993. Pesticides: Historical changes demand ethical choices. Ed. D. Pimentel and H. Lehman. The Pesticide Question: Environment, Economics, and Ethics.
New York: Chapman & Hall, 390-417.

Pimentel, D. 1988. Herbivore population feeding pressure on plant host: Feedback evolution and host conservation. *Oikos* 53:289-302.

Pimentel, D. 1995. Protecting crops. Ed. W.C. Olsen. The Literature of Crop Science. Ithaca, NY: Cornell Univ. Press, 49-66.

Pimentel, D., J. Krummel, D. Gallahan, J. Hough, A. Merrill, I. Schreiner, P. Vittum, F. Koziol, E. Back, D. Yen, and S. Fiance. 1987. Benefits and costs of pesticide use in U.S. food production., *BioScience* 28:778-784.

Pimentel, D., D. Andow, R. Dyson-Hudson, D. Gallahan, S. Jacobson, M. Irish, S. Kroop, A. Moss, I. Schreiner, M. Shepard, T. Thompson, and B. Vinzant. 1980. Environmental and social costs of pesticides: A preliminary assessment. *Oikos* 34:127-140.

Pimentel, D., L.McLaughlin, A. Zepp, B. Latikan, T. Kraus, P. Kleinman, F. Vancini, W.J. Roach, E. Graap, W.S. Keeton, and G. Selig. 1991. Environmental and economic impacts of reducing U.S. agricultural pesticide use. Ed. D. Pimentel. Handbook on Pest Management in Agriculture. Boca Raton: CRC Press, 679-718.

Pimentel, D., H. Acquay, M. Biltonen, P. Rice, M. Silva, J. Nelson, V. Lipner, S. Giordano, A. Horowitz, and M. D'Amore. 1993. Assessment of environmental and economic impacts of pesticide use. D. D. Pimentel and H. Lehman. The Pesticide Question: Environment, Economics and Ethics. New York: Chapman & Hall, 47-84.

Pimentel, D., C. Wilson, C. McCullum, R. Huang, P. Dwen, J. Flack, Q. Tran, T. Saltman, and B. Cliff. 1997. Economic and environmental benefits of biodiversity. *BioScience* 47(11):747-757.

Pimentel, D., O. Bailey, P. Kim, E. Mullaney, J. Calabrese, L. Waltman, F. Nelson and X. Yao. 1998. Will limits of the Earth's resources control human numbers? Submitted to *J. of Environment, Development and Sustainability*.

PN. 1990. Towards a reduction in pesticide use. *Pesticide News* (March).

Potts, G.R. 1986. The Partridge: Pesticides, Predation and Conservation. London: Collins.

Pye, V. and J. Kelley. 1984. The extent of groundwater contamination in the United States. Ed. NAS. Groundwater Contamination. Washington, DC: National Academy of Sciences.

Repetto, R. and S.S. Baliga. 1996. Pesticides and the Immune System: The Public Health Risks. Washington, DC: World Resources Institute.

Reuben, R. 1989. Obstacles to malaria control in India—The human factor. Ed. W.W. Service. Demography and Vector-Borne Disease. Boca Raton: CRC Press, 143-154.

Robinson, W.E., R. Nowogrodzki, and R.A. Morse. 1989. The value of honey bees as pollinators of U.S. crops. Am. Bee J. 129:477-487.

Schuman, S. 1993. Risks of pesticide-related health effects: An epidemiologic approach. Ed. D. Pimentel and H. Lehman. The Pesticide Question: Environment, Economics, and Ethics. New York: Chapman and Hall, 106-125.

Stickel, W.H., L.F. Stickel, R.A. Dryland, and D.L. Hughes. 1984. DDE in birds: Lethal residues and loss rates. Arch. Environ. Contam. and Tox. 13:1-6.

Stone, W.B. and P.B. Gradoni. 1985. Wildlife mortality related to the use of the pesticide diazinon. Northeastern Environ. Sci. 4:30-38.

Storey, D.A. and P.G. Allen. 1993. Economic impact of marine recreational fishing in Massachusetts. N. Am. J. Fish. Man. 13:698-708.

Stringer, A. and C. Lyons. 1974. The effect of benomyl and thiophanate-methyl on earthworm populations in apple orchards. Pesticide Sci. 5:189-196.

Tanda, A.S. 1984. Bee pollination increases yield of 2 interplanted varieties of asiatic cotton (gossyium arboreum l.) (apis cerana indica, apis mellifera, India). Am. Bee J. 124:539-540.

Trumbel, J.T., W. Carson, H. Nakakihara, and V. Voth. 1988. Impact of pesticides for tomato fruitworm (Lepidoptera:Noctuidae) suppression on photosynthesis, yield, and nontarget arthropods in strawberries. J. Econ. Entomol. 81:608-614.

UNEP. 1979. The State of the Environment: Selected Topics—1979. Nairobi: United NationalEnvironment Programme, Governing Council, Seventh Session,

UNEP. 1997. Global Environmental Outlook. Nairobi: United Nations Environment Programme.

USBC. 1994. Statistical Abstract of the United States 1994. Washington, DC: U.S. Bureau of the Census, U. S. Government Printing Office.

USDA. 1989. Agricultural Statistics. Washington, DC: U.S. Department of Agriculture, Government Printing Office.

USDA. 1994. Agricultural Statistics. Washington, DC: U.S. Department of Agriculture, Government Printing Office.

USFWS. 1988. 1985 Survey of Fishing, Hunting, and Wildlife Associated Recreation. U.S. Fish and Wildlife Service. U.S. Dept. of Interior: Washington, DC.

Van der Leeden, F., F.L. Troise, and D.K. Dodd. 1990. The Water Encyclopedia. 2nd ed. Chelsea, MI: Lewis Pub.

Von Rumker, R. and F. Horay. 1974. Farmers' Pesticide Use Decisions and Attitudes on Alternate Crop Protection Methods. Washington, DC: U.S. Environmental Protection Agency.

Ware, G.W. 1983. Reducing pesticide application drift-losses. Tucson: Univ. of Arizona, College of Agriculture, Cooperative Extension.

Ware, G.W., W.P. Cahill, B.J. Esesen, W.C. Kronland, and N.A. Buck. 1975. Pesticide drift deposit efficiency from ground sprays on cotton. J. Econ. Entomol. 68:549-550.

Weiner, B.P. and R.M. Worth. 1972. Insecticides: Household use and respiratory impairment. In Adverse Effects of Common Environmental Pollutants. New York: MSS Information Corporation. 149-151.

White, L. 1967. The historical roots of our ecological crisis. Science 155:1203-1207.

WHO. 1992. Our Planet, Our Health: Report of the WHO Commission on Health and Environment. Geneva: World Health Organization.

WHO. 1996. Micronutrient malnutrition—Half the world's population affected. World Health Organization, 13 Nov 1996, no. 78, 1-4.

Williamson, C.S. 1995. Conserving Europe's bees: Why all the buzz? TREE 10:309-310.

Wills, L.E., B.A. Mullens, and J.D. Mandeville. 1990. Effects of pesticides on filth fly predators coleoptera histeridae staphylinidae acarina macrochelidae uropodiae in caged layer poultry manure. J. Econ. Entomol. 83:451-457.

FALLING LEAVES AND ETHICAL DILEMMAS: AGENT ORANGE IN VIETNAM
Arthur W. Galston

Introduction

Advances in science and technology frequently present us with ethical dilemmas that are difficult to resolve, partly because of their intellectual novelty, but also because society has so little experience coping with the kinds of problems presented by new science. So it was, after the first atomic explosion forced us to confront how to regulate this powerful new force to benefit, rather than to destroy, humankind. So it is now, with our newfound ability to clone mammals and to engineer the genetic makeup of cells and organisms. No field of science, no matter how innocuous it may seem, is exempt from an ability to stir the ethical pot, yielding vexing new problems. What follows here is an account of a major ethical dilemma that emerged from research on green plants. As a result of this work, the botanist, probably the last of the scientific innocents, was unexpectedly catapulted into the same ethical hot pot as other scientific colleagues.

It all started with the United States' entry into World War II. After the attack on Pearl Harbor on December 7, 1941 and the initial rapid advance of Japanese forces over many islands of the Pacific, United States forces began a slow, methodical campaign to regain the lost territories as a prelude to an assault on the home islands of Japan. A typical invasion of a Pacific island began with a massive naval artillery bombardment to neutralize gun emplacements and fixed defenses, after which marine and army personnel, offloaded from transport ships to landing craft, stormed ashore and tried to fight their way inland to engage the enemy. This operation frequently took a horrible toll on our fighting men, since enemy machine guns hidden in the dense jungle vegetation spit out a stream of death-dealing fire. Our military commanders, aware of the terrible price in American lives being exacted for each bit of reconquered territory, began to explore methods to diminish the danger by first removing the dense leafy cover hiding the enemy's firepower. They first thought to accomplish this with new types of artillery shells designed to blast leaves off stems, but found this technique inefficient and impractical. You can knock a tree over with a blast, but it will usually fall with most of its leaves in place.

Eventually, military leaders turned to plant physiologists, specialists whose knowledge of plants, especially of the hormonal control of plant function, permitted them to control at will the shedding (abscission) of leaves. Through grants to botany departments at universities (greeted like manna from heaven by these perennially underfunded institutions!) and through the establishment of specialized research institutes like the Army Chemical Corps laboratory at Fort Detrick in Frederick, Maryland, military agencies catalyzed investigations which led eventually to the development of efficient chemical weapons for the defoliation of trees. Similar operations in the United Kingdom provided some of the basic knowledge and field testing underlying this achievement.

World War II ended before the technology of defoliation had reached the point of providing a reliable military weapon. In the final phases of military action during and immediately after the war, when the British were trying to retain their control of certain Asian territories, they used defoliating chemicals in parts of the Malay peninsula both to minimize the possibility of ambush by clearing important roads, trails and waterways, and also to kill food crops in areas suspected of being designated for insurgent use. This limited activity produced data useful in the design of ever more efficient procedures for the military use of the active compounds.

By the time of the United States' maximum involvement in the Vietnam War in the late 1960s, several herbicides and defoliants were part of our military arsenal. Out of more than a thousand chemicals tested at Fort Detrick and elsewhere during World War II, chlorinated phenoxyacetic acids, which mimicked the action of naturally-occurring plant hormones, emerged as the most effective compounds for defoliation operations. Two of these compounds, named 2,4-D (2,4-dichlorophenoxyacetic acid) and 2,4,5-T (2,4,5-trichlorophenoxyacetic acid), were incorporated into an effective defoliant, called Agent Orange. Vast quantities of this material were sprayed over Vietnam by air. The operation covered an area about two-thirds the size of the state of Massachusetts, enough to make it the largest chemical warfare operation in history.

This campaign raised a number of important ethical questions, with which we are still grappling. First, did these operations violate existing international laws and treaties against the use of chemical weapons? Second, did the sprayed chemicals have lasting deleterious effects on the productivity of the land, the waters and the ecological systems of Vietnam? Were they harmful to animals and to the people inadvertently sprayed, including Vietnamese civilians and military personnel of both nationalities? Did the sprays destined for forests drift appreciably, causing destruction of food crops used for sustenance by non-combatant populations, including women and children? Balancing the benefits and harmful effects of this kind of operation, would we repeat our actions in a future war? These and other troublesome questions will be examined below.

The science behind the chemical defoliation campaign

To understand Agent Orange, we need to understand something about plant hormones, which regulate many of the important life processes of plants. Plant hormones were discovered about 1928,[124] and first chemically identified in the 1930s.[125] In vanishingly small concentrations approximating the parts per billion range, they are able to effect major chemical and structural changes in the plant. They accomplish these prodigious effects through their control of the activity of genes, membranes and other components of the cell, permitting amplification of a chemical signal. If, for example, a hormone is able to activate a previously inactive gene, then each time the gene functions to produce its product (a catalytically active protein), the hormone's effect is multiplied manifold. Similarly, if a hormone regulates the passage of a biologically active molecule from one cellular compartment to another through a membrane, it can greatly affect that compound's activity and metabolism. Hormones use both mechanisms in plants.

At least five plant hormones are now recognized,[126] of which only two, *auxin* and *ethylene*, need concern us here. Auxin, known chemically as 3-indoleacetic acid (IAA), is produced near the tips of stems and flows mainly downward through living cells, including those in the petioles (leaf stalks) attaching the leaf blades to the stems. If the level of auxin flowing into the leaf blade is low and stable, the leaf blade remains attached to the plant by a strong and resilient *abscission layer* at the point of attachment of the petiole to the stem. But if the concentration of auxin rises, either through internal mechanisms or through external application, certain inactive genes are "turned on" and new biochemical reactions occur in the abscission layer. First, a common amino acid called methionine is dismembered into several pieces, including ethylene, a gaseous hormone. Ethylene in turn causes production of the enzyme *cellulase*, but only in the abscission layer. Cellulase is able to digest the cellulosic cell wall that surrounds every plant cell. When such action weakens the abscission layer cell walls sufficiently, the leaf falls from the tree in response to displacement by a breeze, a mechanical blow, or gravity alone.[127] In nature, the abscission of leaves is usually

---

[124] Went, F.W. Wuchsstoff und Wachstum. *Recueil des Travaux Botaniques Neerlandais*, 1928. 25:1-116.

[125] Kogl, F., A.J. Haagen-Smit & H. Erxleben. Uber ein Phytohormon der Zellstreckung. Reindarstellung des Auxins aus menschlichen Hern. IV. *Mitteilung. Ztschr.Physiol.Chem.* 1933. 214:241-261.

[126] Galston, A.W. Life Processes of Plants. Scientific American Library. New York: W.H. Freeman & Co., 1994.

[127] Addicott, F.T. Abscission. Berkeley: University of California Press, 1982.

governed by the relative length of day and night, which in turn controls hormone levels.[128]

Understanding this mechanism has permitted us to control abscission artificially though the use of synthetic molecular mimics of the controlling hormones. In some instances, spraying these compounds facilitates production or harvest of certain crops. For example, to gather ripe cotton bolls by a mechanical harvester instead of hand labor, one must first remove the leaves from the plant, since they tend to clog the machinery. To accomplish this, we first spray the crop with a synthetic auxin 48-72 hours before the contemplated harvest. The leaves then conveniently abscise, permitting mechanical harvest to proceed smoothly. It is not difficult to adapt a similar regimen to the defoliation of tropical trees and bushes typical of the Pacific islands slated for invasion.

The military use of defoliation

The mixture of 2,4-D and 2,4,5-T, designed for use as a aerial spray, and stored until use in 55-gallon barrels bearing an orange stripe, became known to the U.S. Army as Agent Orange. During the Vietnam was, approximately 100 million pounds of this mixture, applied as a 25 lb. per acre spray released over a broad swath of forest by groups of fixed-wing aircraft, were used in planned defoliation operations.[129] This operation, known as "Ranch Hand," had as its jaunty motto, "Only you can prevent a forest." In addition to Agent Orange, formulations of other herbicides were employed for specific purposes, such as the selective killing of rice, but Agent Orange was by far the most widely employed herbicide used in Vietnam.

In the Vietnamese conflict, Agent Orange found use in two major theaters of warfare, the inland hardwood forests and the mangrove communities lining the estuaries of the Mekong delta south and west of Saigon. The forests were sprayed in order to permit aerial interdiction of the transport of men and materiel along what came to be called the "Ho Chi Minh Trail." This 1200 mile long collection of trails, running all the way from Hanoi in the north to Saigon in the south, represented a truly Herculean effort by determined fighters to use human muscle and effort to blunt the overwhelming technical might of a wartime antagonist with much greater wealth and power. Bicycles, strengthened and enlarged by metal frames that permitted them to carry half a ton or more of supplies, were doggedly pushed along this lengthy path by soldiers. The troops

---

[128] Westing, A.H., Ed. Herbicides in War: The Long-term Ecological and Human Consequences. Stockholm International Peace Research Institute. London: Taylor & Francis, 1984.
[129] Vince-Prue, D. Photoperiodism in Plants. Berkshire: McGraw-Hill Book Co. UK, 1975.

and supplies thus ferried economically from the north to the combat areas of the south permitted the continuation of effective military action in the face of fierce bombing operations. Defoliation of forested areas through which the trail ran conferred some military advantages, although, as we shall see below, there were unexpected disadvantages as well.

In general, upland forests sprayed once shed their leaves a few days after spray, then sent out new leaves and branches from previously dormant buds. After a second spray, some trees recovered and some did not, but after a third spray, the majority of trees were killed. This extensive abscission led to nutrient dumping, since minerals were leached quickly from the fallen leaves and mineral-poor forest soils by the abundant rainfall of the region. The killed trees were often replaced by previously dormant or light-limited plants of the forest floor, such as scrub bamboo. Unhappily, some of the most valuable timber trees such as teak (*Tectona grandis*) were wiped out in this way, and the bamboo that frequently replaced these valuable trees had little or no economic value. Dense bamboo regrowth on the forest floor made the resulting thicket even denser and thus more amenable to the concealment of guerrilla-type military operations. To add to the paradox, these thickets could often not be killed by any known herbicide, and thus could be cleared only by fire.

The mangrove communities suffered an even more catastrophic fate. The few species constituting this ecosystem have a remarkable ability to flourish in either the salt water of the sea or the almost fresh water flowing into the estuary from rivers. They stand high above the water, their aerial roots providing a platform from which their leafy branches grow. Mangroves are tremendously important ecologically because many fish and shellfish grow and reproduce in the sheltered microenvironment provided by their filigree of roots. The densely interlaced leaves of these communities unfortunately also provided a sheltered environment for Vietcong fighters who ambushed American patrol vessels guarding the security of the estuaries. Thus, a fateful decision was made to spray Agent Orange over the mangroves. The tragic result, which could not have been anticipated, was almost total destruction of this specialized vegetation community, which has not yet completely recovered, more than two decades after the end of the war. Death of the mangroves was accompanied by concomitant loss of the fish and shellfish communities normally harbored among their roots. Since fish constitute the major source of protein in the Vietnamese diet, and are in addition the basis for the pungent *nuoc mam* sauce into which almost all Vietnamese food is dipped, the aerial spraying of the mangroves had the unanticipated effect of inducing serious deficiencies in the civilian population. This was certainly not an effect American military strategists had anticipated or wanted; unfortunately, it was not the only surprising consequence of Operation Ranch Hand.

Toxicology of Agent Orange

The safe aerial dissemination of toxic materials, whether designed to act against plants or animals, requires adherence to practices that restrict toxin deposition to the intended target area. Thus, the altitude of the airplanes, their velocity, the size of the aerosol droplets used, wind velocity and direction, and terrain all play a role in the calculation of safe procedure. Where such operations are carried out over peaceful agricultural fields in the United States, and where a mistake in toxin deposition could easily lead to an expensive lawsuit, both the planners of the operation and pilots carrying it out are moved to follow safety rules meticulously. But a similar operation in a distant land, under military conditions involving some danger from unfriendly ground fire, would almost certainly be carried out with much less regard for the niceties of safety rules. So it is not surprising that planes spraying Agent Orange sometimes flew higher and faster than they should have to avoid enemy ground fire, and that details such as wind speed would have been overlooked in order to cover a critical area in a timely fashion. The result was that considerable defoliation occurred in unintended areas relatively close to targets; such areas included numerous fruit trees like papaya and jackfruit planted alongside peasant houses in villages, broadleaved agricultural crops such as sweet potatoes, and even a Michelin rubber plantation over the border in Cambodia. The adverse economic, nutritional and public health consequences of such unintended sprays have never been precisely estimated, but it is certain to have run into the millions of dollars of value.

But there was another, even more serious, unintended effect of the Agent Orange spray, caused by a group of impurities produced during synthesis of 2,4,5-T. This compound is synthesized by combining 2,4,5-trichlorophenol with a modified acetic acid under alkaline conditions. But during synthesis at elevated temperatures designed to make the reaction proceed more quickly, an unwanted side reaction occurs. Two molecules of the chlorinated phenols react with each other to form a tricyclic planar compound with four chlorine atoms at the periphery of the plane. Such compounds are able to insert themselves into the groove between the two complementary chains of the duplex DNA molecule, thereby interfering with basic replicative processes essential to the cell. These inadvertently produced dioxins, such as 2,3,7.8-tetrachloro-para-dibenzodioxin (TCDD), turned out to be extremely toxic to both humans and animals.

The possible harmful effects of these dioxin contaminants on American servicemen subjected to aerial spray in Vietnam formed the basis for a historic class action lawsuit against the chemical companies which manufactured Agent Orange for the government. This mass tort trial was eventually settled by the payment of many millions of dollars to the veterans. There has still been no campaign to analyze possible toxic effects on the Vietnamese population. This is both paradoxical and tragic, since the most obvious and harmful effect of the dioxins is the production of serious fetal malformations (teratogenies) during critical periods of embryonic development *in utero*. Pregnant Vietnamese women and their unborn children were clearly at risk.

The campaign against the use of defoliants in Vietnam

Suspicions about possible toxic effects of chemical herbicides and defoliants on humans led to my interest and involvement with the fight against Agent Orange. As a graduate student at the University of Illinois during the early years of World War II, I was seeking a chemical that might speed up the flowering of certain strains of soybeans, then being introduced into American agriculture from China. I was fortunate in finding such a compound in 2,3,5-triiodobenzoic acid (TIBA), a recently discovered weak auxin. I noted in my Ph.D. thesis in 1943[130] that to receive this desirable effect, one had to carefully control the concentration of TIBA in the spray, for higher concentrations were extremely toxic, producing malformations of buds and abscission of mature leaves. Having defended my thesis successfully in the spring of 1943, I turned for the next three years to war-related work at the California Institute of Technology and then to military service.

About a year after my discharge from the U.S. Navy in the spring of 1946, after I had returned to Cal Tech, I was visited by Drs. A. Geoffrey Norman and Robert L. Weintraub, two senior plant physiologists from Fort Detrick, who informed me that my thesis work on the defoliation effects of TIBA had served as a model for further investigation on defoliants. Eventually, TIBA was discarded in favor of 2,4-D and 2,4,5-T, the active components of Agent Orange, but it provided the scientific and emotional link that compelled my involvement in opposition to the massive spraying of these compounds during the Vietnam war.

On December 21, 1965, the *New York Times* reported that United States military forces in Vietnam were employing aerially-dispersed herbicides to defoliate forests and, in localized areas, to deprive guerrillas of food. Agent Orange was the principal material employed in selective killing of rice. Alarmed by the possible public health and ecological consequences of the massive use of such materials, whose toxicology I knew was virtually untested, I prepared a resolution questioning such use, for presentation to the August 1966 annual meeting

---

[130] Galston, A.W. 1943. The Physiology of Flowering, with Special Reference to Floral Initiation in Soybeans. Ph.D. Thesis. Univ. of Illinois, Urbana.

of the American Society of Plant Physiologists. To my chagrin, the Executive Committee failed to recommend this resolution for presentation to the meeting as official business of the Society. When I insisted on reading the resolution from the floor, I received support from a portion of the membership, and when I circulated copies, more than a dozen of my colleagues signed the document, which we sent as a petition to President Lyndon Johnson at the White House. The letter said in part:

> The undersigned plant physiologists wish to make known to you their serious misgivings concerning the alleged use of chemical herbicides for the destruction of food crops and for defoliation operations in Vietnam. The use of such agents by United States forces was reported in the *New York Times* of December 21, 1965 and has never been denied by the Administration or by the leaders of our military operations. Our deliberations and our statements below are based on the assumption that this published report is true.

> We would assert in the first place that even the most specific herbicides known do not affect only a single type of plant. Thus, a chemical designed to defoliate trees might also be expected to have some side effects on other plants, including food crops. Secondly, the persistence of some of these chemicals in soil is such that productive agriculture may be prevented for some years. Thirdly, the toxicology of some herbicides is such that one cannot assert that there are no deleterious effects on human and domestic animal populations. It is safe to say that massive use of chemical herbicides can upset the ecology of an entire region, and in the absence of more definite information, such an upset could be catastrophic.

> Even if we assume that our military leaders have selected reasonably specific anti-rice herbicides, nontoxic to humans or to domestic animals, for use in Vietnam, we must still be concerned with the effects of large-scale food deprivation on a mixed civilian-military population. As Professor Jean Mayer of the Harvard School of Public Health pointed out in a letter to *Science* on April 15, 1966, the first and major victims of any food shortage or famine, caused by whatever agent, are inevitably children, especially those under five. This results mainly from their special nutritional needs and vulnerability to stress. Thus, the effect of our use of chemical herbicides may be to starve children and others in the population whom we least wish to harm.

We received a reply dated September 28, 1966 from then Undersecretary of State Dixon Donnelley, which said in part:

> Chemical herbicides are being used in Vietnam to clear jungle growth and to reduce the hazards of ambush by Viet Cong forces. These chemicals are used extensively in most countries by both the Free World and the Communist Bloc for selective control of undesirable vegetation. They are not harmful to people, animals, soil or water.

> The elimination of leaves and brush in jungle areas enables our military forces, both on the ground and in the air, to spot the Viet Cong and to follow their movements, and to also avoid ambushes.

> Destruction of food crops is undertaken only in remote and thinly populated areas under Viet Cong control, and where significant denial of food supplies can be effected by such destruction. This is done because in the Viet Cong redoubt areas food is as important to the Viet Cong as weapons. Civilians or non-combatants are warned of such action in advance. They are asked to leave the area and are provided food and good treatment by the Government of Viet-Nam in resettlement areas.

This exchange prompted me to investigate what international law or convention had to say about the use of chemical weapons to damage or kill plants. The Geneva Gas Protocol of June 17, 1925 states in part:

> Whereas the use in war of asphyxiating, poisonous or other gases, and of all analogous liquids, materials or devices, has been justly condemned by the general opinion of the civilized world; and,

> Whereas the prohibition of such use has been declared in Treaties to which the majority of Powers of the world are Parties; and

> To the end that this prohibition shall be universally accepted as a part of International Law, binding alike the conscience and the practice of nations;

> Declare that the High Contracting Parties, so far as they are not already Parties to Treaties prohibiting such use, accept this prohibition, agree to extend this prohibition to the use of bacteriological methods of warfare and agree to be

bound as between themselves according to the terms of this declaration.

Although this treaty was drawn up largely by U.S. Secretary of State Frank Kellogg, it was never ratified by our isolationist-minded Congress, then in the midst of a post-World War I rejection of the League of Nations and other Wilsonian projects. The U.S. did, however, sign a United Nations Resolution of December 5, 1966, which called on signatories to observe 1925 protocol. Despite our commitment not to use "asphyxiating, poisonous or other gases," our forces in Vietnam used more than 14 million lbs. of so-called "riot control gas" (o-chloromalononitrile, or CS) against Vietnamese civilians and soldiers. To explain the apparent contradiction between our practice and promises as a signatory to the United Nations resolution, our Ambassador to the United Nations James Nabrit said:

> The Geneva Protocol of 1925 prohibits the use in war of asphyxiating and poisonous gas, and other similar gases and liquids with equally deadly effect. It was framed to meet the horrors of poison gas warfare in the first World War, and was intended to reduce suffering by prohibiting the use of poisonous gases, such as mustard gas and phosgene, It does not apply to all gases. It would be unreasonable to contend that any rule of international law prohibits the use in combat against an enemy for humanitarian purposes of agents that governments around the world commonly use to control riots by their own people. Similarly, the protocol does not apply to herbicides, which involve the same chemicals and have the same effects as those used domestically in the United States, Soviet Union, and many other countries to control weeds and other unwanted vegetation.

The description by our government of its use of chemical warfare agents in Vietnam led to numerous confrontations between scientists and the military. Sparked by Professors A.H. Westing of Windham College and E.W. Pfeiffer of the University of Montana, the American Association for the Advancement of Science directed inquiries to the responsible military agencies, and later appropriated $80,000 to fund an Herbicide Assessment Commission, headed by Professor M.S. Meselson of Harvard University. As a consequence of this Commission, the Department of Defense was moved to sponsor a literature research project on antiplant chemicals, carried out by the Midwest Research Institute. Later on, prodded by persistent reports of the toxicological effects of sprayed defoliants, the DoD awarded laboratory research contracts to the Bionetics Research Labs of Litton Industries for investigation of their possible mutagenic, carcinogenic and teratogenic effects.[131] In a report by Bionetics Laboratories to the Surgeon General of the U.S., Bionetics revealed that the phenoxyacetic acids, especially 2,4,5-T, were teratogenic when fed to pregnant rodents in the range of about 200 parts per million of body weight. Human toxicity could be logically inferred from these data.

One mysterious aspect of these toxicity studies was that 2,4,5-T from different sources consistently displayed different levels of toxicity. This was not surprising, since the DoD had recruited a variety of manufacturers, including Dow, Monsanto, Dupont, Hercules and Thompson to fill its prodigious orders for Agent Orange. Systematic examination of this problem led Meselson and others to conclude that a trace contaminant, present in different concentrations in different samples of the herbicide preparations, accounted for the toxicity, not the 2,4,5-T itself. Careful chemical work resulted in the discovery and isolation of several members of a family of contaminants, the dioxins. One particular dioxin, TCDD, has turned out to be one of the most toxic organic compounds known, producing teratogenic effects in rats when given in the parts per billion (ppb) range and in primates in the parts per trillion (ppt) range.[132] The effects ranged from cleft palate and cystic kidneys to cancers and stillbirths, in a dose-dependent manner.

Through an accident of history, we were able to present this dramatic information directly to the highest councils of the Nixon administration in the latter part of 1969. President Nixon's Science Adviser was the distinguished physicist Lee A. DuBridge, who had been president of the California Institute of Technology in Pasadena before accepting this Washington assignment. Meselson had been a graduate student in Chemistry under Linus Pauling, and having gained some research notoriety since his CalTech days and a Professorship at Harvard, was well known to DuBridge. I was also known to DuBridge, because I had been an Associate Professor of Biology at the Institute, and had also been an anti-McCarthy and anti-HUAC (House Unamerican Activities Committee) activist during the days when this organization was hounding Pauling for circulating the antinuclear Stockholm petition. (My activities were described luridly in the right-wing newspaper, the Pasadena *Independent*, and DuBridge once had to inform me that these activities were possibly jeopardizing my scientific career.)

An explanatory telephone call from Meselson to DuBridge set up a meeting at the Old Executive Office Building at which the relevant data

---

[131] Teratogeny refers to abnormal development, and the production of teratomas, or monstrous growths.

[132] Allen, J.R., D.A. Busotti, J.P van Miller, L.J. Abrahamson and J.J. Lalich. Morphological changes in monkeys consuming a diet containing low levels of 2,3.7.8-tetrachlorodibenzo-p-dioxin. *Food Cosmet. Toxicol.* 15:401-410. 1977.

were presented to DuBridge and to some of the military scientific advisers who had authorized use of the Agent Orange spray. The result was DuBridge's recommendation to Nixon that the spray operation be terminated. This resulted in a halt late in 1970, after some months of delay caused by field commanders reluctant to "waste" their stockpiles of Agent Orange before the ban went into effect. Since the war in Vietnam did not end until about five years later, our activities had thus spared the Vietnamese people and countryside from this potentially noxious spray over half a decade.

"Agent Orange on Trial"[133]

A book with this title, written in 1986 by Peter Schuck, a Yale Law School colleague, described one of the largest mass tort trials in history. The trial arose through a confluence of separate actions filed by otherwise healthy American Vietnam war veterans who came down suddenly and mysteriously with a variety of illnesses ranging from cancer to peripheral neuropathy and skin diseases. When information about the deleterious effects of dioxins began to appear in the public press, and when the dioxins were connected with Agent Orange, groups of veterans, both with and without the kinds of health problems attributable to dioxins, began to explore the possibilities of a class action law suit against the government and the chemical companies that supplied the government with the defoliants it had requested. In this manner, the original lawsuit filed by a single individual, Paul Reutershan, was converted into a 100,000-member organization called Agent Orange Victims International, and then into a massive legal action involving thousands. Although Paul Reutershan died of cancer, AOVI carried on the struggle for compensation under the leadership of a disillusioned former GI named Frank McCarthy, who was also embittered by the American public's disdain for Vietnam veterans. After overcoming some initial resistance, McCarthy was able to interest a feisty lawyer named Victor Yannacone in pursuing the class action suit. Yannacone converted the original Reutershan complaint into a class action suit in January of 1979, but it was not until almost seven years later, after Yannacone was no longer part of the AOVI legal team, that a settlement was reached.

The veterans had originally sought to sue the government as well as the chemical companies, but this possibility was denied them following a ruling by Judge George C. Pratt under the Federal Tort Claims Act of 1946. The veterans' case against the chemical companies dragged on for years under a fact-finding and data gathering regime devised and supervised by the methodical Judge Pratt. When Pratt's

---

[133] Schuck, P.H. Agent Orange on Trial: Mass Toxic Disasters in the Courts. Cambridge: The Belknap Press of Harvard University Press, 1986.

elevation to the Court of Appeals led to his replacement by the more dynamic Judge Jack Weinstein, the court case took a new turn that led to its resolution in a fairly brief span of time. Weinstein was able to persuade both sides that a settlement, rather than a trial, was the most desirable route to follow. After many negotiations, with dissident opinions from minority groups on both sides, the chemical companies agreed, without admitting culpability, to contribute $180 million plus interest to a fund that would be used in part to compensate those litigants with strong claims.

Many of the veterans had hoped for a larger settlement, but were ultimately moved to accept the Weinstein formula because of the obvious difficulty of proving that any one cancer or neuropathy was in fact caused by Agent Orange rather than some other agent or to some "spontaneous" event. To this day, it remains controversial whether sprayed dioxins actually caused the ailments afflicting the veterans. Even if dioxin were proved culpable, Dow and other chemical companies presented convincing evidence that these substances could arise from a multiplicity of products and technologies, including incineration of wastes in disposal plants and, in general, fires of any type. It would thus have been virtually impossible for any individual or group of veterans to prove that dioxin in Agent Orange was the source of their pathology without a lengthy and expensive trial involving many people in control and experimental groups randomized and matched so as to meet demanding requirements for statistical significance.

Of the medically important effects known to be due to dioxin, the two most convincing are chloracne, a severe skin ailment, and teratogeny, or malformation of embryos *in utero*. Almost all of the veterans were, of course, males, so potential teratogenic effects were immediately ruled out of this population. But there were many pregnant women in the heavily sprayed Vietnamese villages, and simple calculation based on known spray rates and reasonable consumption of water and food revealed that some of them could have consumed presumed teratogenic doses of TCDD through mere use of drinking water. The Vietnamese had claimed consistently that the occurrence of teratogeny and stillbirths had risen significantly in locales that were heavily sprayed with defoliant, and Dr. Ton That Tung showed me his data and collections of teratogenic embryos at the Viet Duc hospital in the early 1970s. American scientists refused to credit these claims as long as there were no diplomatic relations between the two countries, and because it would be difficult to establish whether such claims were factually accurate or concocted to exact retribution from a powerful enemy. That situation is changing now that the war has receded into the past and official diplomatic ties have been established. Still, some medical investigators from the United States have faced difficulties in their attempts to gather credible data that would satisfy American colleagues.

Ethical questions raised by the military use of defoliants

War is by its very nature violent and destructive, yet there have always been rules designed to mitigate its most horrible aspects. Just as there are rules of warfare governing the treatment of prisoners and civilians, there are other conventions, many unwritten, proscribing such environmental practices as the salting of wells, the deliberate spreading of disease germs and a "scorched earth" policy involving the wrecking of food supplies and agricultural facilities that support the civilian population. How does the use of antiplant chemicals fit into these proscriptions? Aside from the question of the possible violation of international treaties against chemical warfare, several important ethical questions are raised by the use of Agent Orange in Vietnam.

One of these concerns the issue of whether our mass chemical attack on the ecology of Vietnam can be called *ecocide*. This term was coined to evoke the specter of the parallel crime of genocide, justly condemned after the Nuremberg trials, as the deliberate extermination of members of an ethnic group solely because of their ethnicity. (This problem is still very much with us; we need but recall the massacre of ethnic innocents in Bosnia, Rwanda and Kosovo.) The acknowledgement of genocide as a crime against humanity raises a parallel question regarding the environment. Should it also be considered a crime to alter the ecosystem upon which the livelihood of a group depends so that the land is no longer able to support that group? Claiborne Pell, then Senator from Rhode Island, advanced a proposal in 1972 which would have banned any environmental or geophysical modification as a weapon of war. This statement later became a "sense of the Senate."[134] Several authors[135] and governments have also proposed legislation that would ban *ecocide*, any military action designed to modify terrain or weather so as to adversely affect the stability or productivity of the ecosystem.

Although the defoliation operation in Vietnam did not permanently destroy the productivity of the ecosystem, considerable damage flowed from "nutrient dumping" caused by premature abscission of leaves, from increased soil erosion and flooding during increased runoff of water, the destruction of the aquatic mangrove fish habitat, the deposition of long-lived toxic residues in the soil, and finally, the

---

[134] Pell, C. Senate Resolution No. 281, 92nd Congress. U.S. Congressional Record, Washington, 118:8871-8873; 9277; 13386, 1972. Also Senate Resolution No. 71, 93rd Congress. U.S. Congressional Record, Washington, 119:22303-22305, 1973.

[135] Falk, R.A. Environmental warfare and ecocide. *Bulletin of Peace Proposals*, Oslo 4:1-17, 1973.

substitution of "junk" vegetation like scrub bamboo for valuable species like teak.[136] While this damage has not been accurately assessed and may never be, the sum is certain to run into many millions of dollars.

Although deliberate destruction of rice and other food crops through the use of cacodylic acid and other toxic substances was not a major activity of Operation Ranch Hand, there were some unfortunate effects of the limited rice-killing operations. On several occasions, our pilots spotted plantations of upland rice in peripheral regions assumed to be dominated by guerrilla fighters with loyalties to the Vietcong. They asked for and received authorization to spray these areas and did so effectively, destroying the crop prior to harvest. Only later was it found that these crops were the major food sources for the indigenous hill people, including the H'mong, many of whom were militarily allied with our forces. Following this crop loss due to "a poisonous rain from the sky," many of these animistic hill people reluctantly left their ancestral homelands for resettlement in guarded lowland encampments. Few returned to their original homes, and some have even resettled in the United States. In this instance, at least, an herbicidal spray has caused the extinction of the life style of a significant number of minority ethnic people.

Natural history surveys of the battered landscape of Vietnam have also revealed the probable extinction of several species of animals and plants, especially in the defoliated areas. Most of these species are economically important, but in view of the current attention being given the importance of the preservation of biodiversity and the interconnectedness of the various components of all ecosystems, any such loss is, of course, unfortunate.

Following the dropping of the atomic bombs in Japan that ended World War II, the victorious United States forces organized an Atomic Bomb Casualty Committee which cared for the survivors while gathering data on radiation damage to affected individuals. But in Vietnam, U.S. forces did not emerge victorious, and assumed no responsibility, following the end of hostilities, for repairing any of the damage to people or ecosystems. So the damage remains uncompensated, while time threatens to obliterate the specific event that caused the disturbances.

A few general conclusions

To sum up, the herbicidal defoliation campaign in Vietnam produced certain militarily advantageous results which may, in the end,

---

[136] Comite National D'Investigation des Consequences de la Guerre Chimique Americain au Viet Nam. Symposium International sur les Herbicides et Defoliants Employes dans la Guerre: Les Effets a Long Terme sur l'Homme et la Nature. Hanoi, 1983.

have justified their use in a bitterly contested war. Defoliation also caused much unintended damage to people, livestock, crops and ecosystems containing wild plants and animals, as well as to soils, rivers, local climate and even entire cultures. The calculation of the relative costs and benefits ought to make it unlikely that any such tactic will be employed in military operations in the future.

ANIMAL MATTERS
Strachan Donnelley

Some years ago the philosopher Mary Midgley wrote <u>Animals and Why They Matter</u>.[1] The book speaks eloquently for itself, and despite much philosophical, theological and ethical exploration of humans' relations to and use of animals in laboratory research and elsewhere, the central question remains very much alive. Why do animals matter? The issue remains fundamentally important because we human beings and the character and goodness of our lives are centrally implicated. Our attitude towards animals and wider animate life says much about ourselves, who we are, and our basic human and moral stance towards the world.

Moral concerns for animals in laboratory research settings are nothing new. Fundamental moral issues of pain, suffering, distress, death and disrespect have long been recognized by Animal Care and Use Committees and researchers in guidelines for research and laboratory care settings.[2] Yet our basic attitudes and moral stances toward animals remain a crucial and ongoing issue, for this is what motivates our moral behavior and concern towards animals—how we follow guidelines and what they are or ought to be. Thus, I want to reopen the discussion of how and why animals matter and re-ask that will-of-the-wisp question: Can we get an overall framework of thought or world view to orient ourselves morally and help guide our decision making on animal use with respect to institutional policies and individual research protocols?

Animals matter to us within particular contexts of human interests, purposes, and concerns, frameworks of though and action, and moral landscapes. These contexts are many, as many (perhaps) as our various contexts of though and action. Thus my strategy here is to explore our encounters with animals in various settings—laboratories, homes, and the wild—and then ask whether these various encounters suggest overall philosophical reflections on why animals matter to us in fundamentally important ways—ways that might not be immediately obvious to us in laboratory settings. Finally, I want to explore specifically what a robust philosophy of organism or organic life might have to say on "animal matters."

Animal encounters

First consider encounters with animals in the scientist's laboratory. Who or what is the animal, say a mouse or a frog, in

---

[1] Midgley, M. 1983. <u>Animals and Why They Matter</u>. Athens: University of Georgia Press.

[2] Donnelley S., Nolan K., Eds. 1990. Animals, Science and Ethics. Special Supplement, *The Hastings Center Report*, 21:3: 1-32.

laboratory research? First and foremost, it is an object of physiological or behavioral inquiry—that is how it primarily "matters." For the practicing scientist, it is an object of scientific curiosity. In the heat of scientific activity, the animal is essentially a physiological or behavioral system unconnected with the environing world. It is an isolated representative or model of physiological, behavioral or animal (biological) existence as such.

The practicing scientist is subjectively motivated by scientific interests, and his or her mind is filled with reigning scientific paradigms and frameworks of thought—whether Galilean-Cartesian materialism, Watsonian genetics, Skinnerian behaviorism, or something else. In fact, these dominant scientific interests and conceptions crucially determine the encounter and what the animal is for the scientist. The animals and human scientists are embedded within an overarching scientific world view which includes its own particular ethos and mores. If the science is materialist and mechanistic, the scientist observes and acts upon what is taken to be a mere material mechanism. Science by definition is universal or abstractly general, and as such the encounter is only with a faceless instance of animate being. The laboratory scientist does not encounter an individual animal in its ordinary worldly context and relations with other individual animals—at least not in his or her role as laboratory scientist or technician. Here the animal is an isolated "it," an object, and not a living worldly subject. This is constitutive of science's limited aspiration of winning universal, objective and empirically verifiable knowledge of biological nature.

All of this significantly changes when the scientist leaves the lab and goes home, where I, for one, am on more familiar ground. When I arrive home from a day of administration and bioethical reflection, there is Nasti, our aging black Labrador retriever. Galilean science, Skinnerian behaviorism, and Kantian moral philosophy are the farthest things from my mind. I want to know whether Nasti has been out lately and whether he has gotten into the garbage. He is no faceless "it" or scientific model for me, but an all too familiar individual. He is lovable, difficult, and annoying, just like the other members of our household, which includes five daughters. Actually, I rarely think of Nasti explicitly in the framework of thought of "animal" or biology. He is just Nasti, one with whom I must live and deal, day in and day out. Nasti does not carry the same weight with me as my wife and daughters, and we often ignore him while pursuing our more humanly immediate interests. Nevertheless, Nasti gets similar attention and care and is definitely an integral member of the lively community that is our home. Moreover, there are times when Nasti and I would rather be off by ourselves, away from the others, whether cross-country skiing along a river or wrestling in the grass.

In short, Nasti exists within the framework of thought and feeling of a familiar living individual, with his own personal character.

Encounters with domestic pets are very different from scientists' involvements with laboratory animals. In the two contexts, the animals exist and matter for human beings in decidedly different ways, and the human actors are animated by dissimilar interest, feelings and thoughts. Both the humans and the animals exist within different webs of human meanings and values. Central to the world of homes and families are the values of care, concern and responsibility for one another, and pets fit integrally into this world. Thus our feelings of concrete responsibility for the care of domestic pets are characteristically different than for laboratory animals. Animals who are experienced as uniquely individual and as "members" of human families have a stronger and more complex ethical hold on us. As they have come to matter importantly to us, so we have come to matter essentially to them. Without our family or other humans to provide care and feeding, Nasti is left in the lurch, unable to fend for himself. He has always and more or less happily lived within the walls of our home, irrespective of walks in the woods. Involvements with most laboratory animals, under the dominant umbrella of scientific meanings and values, are characteristically less personal and intimate. These animals, save perhaps for certain higher mammalian individuals, matter differently and are less dependent on interactions with human others for their own individual well-being. They require what it takes to flourish as laboratory animals, not as members of family households.

There are numerous other philosophically and ethically significant experiences of animals. In particular, there are encounters with animals in the wild, outside the "human city." This is a particularly important species of encounter, in that the animals again matter differently and can play a decidedly different role in determining the terms of the human experience and the particular context of web of meanings and values. Human interests and activities become more and more attuned to animals and animate beings within the natural world. This is a crucial way in which we culturally "civilized ones" regenerate and recreate ourselves bodily, artistically, scientifically, philosophically and ethically. Here is the issue of the vital interplay of the human city (human cultural life) and the natural, animate world.

Encounters with wild animals begin close to nature. While cooking weekend breakfasts, I look out a country kitchen window at birds around a feeder—chickadees, blue jays, cardinals, woodpeckers, doves, sparrows, red-wing blackbirds, and numerous but for me nameless others. I am drawn by the diverse bird species and by the antagonisms, cooperative efforts and natively instituted pecking orders. I may, in passing, be provoked to think of Darwin, artists, or some ethological tract, for here is the kind of primary worldly experience that animates such human explorers. Despite my having enticed the birds onto the back lawn with a feeder, they come and leave on their own terms and as their own creatures, familiar yet other, independent of me. They are what they are in their own being and can well get along without me. My birdseed is gratis, a brief respite in their naturally rigorous lives. They are beyond my direct human responsibilities and dominantly human-centered frameworks of thought.

As I peer out the window, the birds are in the world abroad, undisturbed. I am a mere spectator—a "spectator-bird," to borrow Wallace Stegner's term. I let them be. Correspondingly, I am more or less free to make of them what I will, elaborating my own web of thoughts, meanings and values. The natural spectator in me may be lured out-of-doors in search of other animals in wilder habitats. I may remain an amateur observer of the animal realm or imagine what it might be to become a professional ethologist, legitimizing my native curiosity by scientific inquiry. Whatever the case, this arguably is not our most richly complex and primordial encounter with wild animals. There are further conceptions, meanings, and values to be uncovered. Here the spectating and thinking human being stands over and against the naturally active animals across a divide constituted by disinterested looking. There is something crucial missing about animal and human existence and the way life fully matters to us. The human being has not engaged, or been engaged by, the animals in a single sphere of dynamic interaction. The human individual has not radically discovered its own animality and natural mode of organic being, its primary status in the natural scheme of things. Such firsthand revelations are left to other human encounters with wild animals—for example, those in hunting and fishing.

I will give an example. Each June I go fishing during a mayfly hatch on a remote pond in northern Wisconsin. Other actors in this annual event (a natural high holiday) include my daughter Inanna, red-wing blackbirds and swallows, brown trout, and the mayflies. Typically the pond is still, or there is a slight breeze. It is dusk with a red-orange sun setting behind a blackening forest of evergreen trees. The air is cool, and there are sounds only of birds and mosquitoes. A swallow leaves its perch on a dead stump in the middle of the pond and dives through the air. Immediately it is joined by other birds. Inanna and I quickly row over to the birds. On the water's surface emerge a host of dun-colored mayflies. The surface is broken by a swirl, and a mayfly disappears.

The dusk deepens, with only the faint light of the horizon. Inanna and I hear slurps of feeding trout all around us. We cast our flies to the sounds. Occasionally there is a decent cast, and we dimly see the dry fly riding high on the water. There is a swirl and a sudden, strong pull on the rod. We fight the fish in the dark, trying to keep it away from submerged logs. The fish is lost, or we land it—a cool, smooth, fat-bellied brown trout. We throw it back or put it in the boat's live-box and, along with the blackbirds, go home, leaving the other trout to continue their feast well into the night.

There are several things to note about this human involvement in a mayfly hatch. First, it is the natural and animal world that importantly sets the terms of the human experience and determines reflections on its meanings and values. The concerns and preoccupations of the "human city" are left far behind. Secondly, a human being enters the realm of action as only one among several actors, and the actors are in a fundamental sense on equal terms. All are interlocked in a single dynamic realm, involved in characteristic living activities, whether seeking prey, avoiding predators, or preparing for mating. All this takes place within a community of interaction.

Being a participant actor within, rather than a spectator-observer without, decisively transforms the human experience. The animals are seen in a different and more complex, realistic light. The swallow in this context is no mere physiological and behavioral system or scientific object. Nor is it an object of disinterested curiosity. It is a living individual, an animal subject actively involved in its own world. The swallow patiently and attentively awaits the anticipated mayflies, an unwitting lookout for the parasitic human fishermen. The brown trout are wily, animate others lurking in the depths, strangely beautiful, intricately patterned and colored. It takes the mayflies to bring the trout to the surface and to cast their natural caution to the winds. The mayflies themselves are shapely and delicate emissaries from the (to us) mysterious insect kingdom, the focus of this complex and compelling natural drama, even while they are having their one day in the sun, after a year as nymphs in the muddy bottom of the pond.

In short, these animals are not objects, whether of science or everyday curiosity. Nor are they familiar pets. They are living, individual, and interconnected wild presences, emerging out of, while remaining in, nature. By their own wild otherness and our firsthand interactions with them, they vividly confront us with our own existence as living organisms and shock us back from the provinces of the human city to our place within the wider natural scheme of things. By their own animate being, they force us to probe radically the nature of our own organic being and to question the natural world and its ultimate meaning, values and goodness—why we, animals and nature matter, and matter together.

Two things stand out forcefully: the dynamic interconnection or interaction of animals and their own individual liveliness or being. In the pursuit of philosophical interpretation and ethical responsibility, we need to emphasize both dynamic interconnectedness and individuality, for they bear significantly on questions of the moral status of humans, animals and nature.

Philosophical reflections

These descriptions of animal encounters in different contexts, how and why animals matter, have not been philosophically innocent. They involve humanly reflective interpretations as well as descriptions of ourselves, animals and nature. For example, in the description of the mayfly hatch you may have noted not-too-distant echoes of Darwinian evolutionary biology and ecology, as well as nature poets and philosophers. This cultural overlay is no doubt inevitable for us humans and need not be pernicious if explicitly recognized. Actually, it is the interaction or intermingling of our firsthand experiences and culturally influenced reflections which allows us to come to understand and explicitly value ourselves, animals and nature more or less adequately.

In pursuing how and why animals most fully matter to us and the moral stance of citizens and researchers alike to animals in laboratory research, I want to deepen this philosophical reflection upon animals in the wild.

Darwin. I want to start with Darwin, as interpreted by Ernst Mayr, since arguably, theories of evolutionary biology and ecology must fit into any philosophy of nature and human life that we might honestly embrace.[3] In a nutshell, Darwin's doctrine of evolution involves the evolution of all life and organic species from a common origin via genetic and behavioral variation and natural selection. For our purposes, there are certain crucial aspects of Darwin's theory that merit emphasis. There is the famous demise of the Grand Design and Designer of Nature (cosmic teleology). Nature evolves its own forms of order—organismic and ecological—through dynamic evolutionary and ecological processes ruled or governed by no one outside of the system. Just as importantly, evolutionary thought spells the end of Newtonian causal determinism— we can no longer operate with a purely linear model dictated by the hegemony of efficient causation, with antecedents solely determining consequents. Rather, organic events—whether genetic, organismal, or ecological—are multi-caused, involving multi-spatial and temporal scales. Systems thinking now enters as a dominant mode of scientific and philosophical thought.

Perhaps even more importantly, Darwinian biology spells the end of typological or essentialist thinking as a fundamental mode of interpreting nature. We no longer have merely the species "dog," "horse" and "human being" but have moved to populational models of thinking. Individual organisms live and interact within interbreeding populations (as well as within the wider world). Each individual organism is most often genetically and phenotypically unique, different from all the others. These fundamental themes of individuality, particularity and diversity are the backbone of the evolutionary story and hold for populations,

---

[3] Mayr, E. 1991. <u>One Long Argument</u>. Cambridge: Harvard University Press.

communities of organisms, ecosystems and bioregions as well. Moreover, these themes are conjoined with equally fundamental ideas of the importance of historical context, dynamism and contingency—the direct or indirect interconnections of all forms of life in the unique "story" that is the historical evolution of life on this planet.

Our interpretive philosophical ears should perk up. The fundamental lineaments of the evolutionary, ecological account seem interpretively a lot closer to our experiences of mayfly hatches and other animals in the wild than they do to our encounters with animals in scientific laboratories. Remember that in labs the individuality and particularity of animals is played down, if not eclipsed altogether, and universality and sameness are played up, in an emphasis on the constant laws of nature and organic being. Here Newton, not Darwin, seems to reign supreme. In laboratories, something important is often, indeed characteristically, left out of account—precisely this fundamental world view and message of (Darwinian) evolutionary, ecological nature. From a philosophical or moral perspective, is our laboratory science too partial or abstract, leaving too much out of account?

This is not all. Is something important left out of account in Darwinism, in scientific evolutionary and ecological theory itself? Does evolutionary biology fully capture or interpret the mayfly hatch and why, finally, animals matter to us? In the biological accounts, we have individual animals (including ourselves) interconnected in dynamic historical contexts. But do these accounts acknowledge their particular individual liveliness, their subjective animal agency, as opposed to an interpretation as mere biological machines, energy exchangers, or genetic reproducers? Where are the living individual animals, with all that they philosophically, spiritually and ethically connote? Where is Nasti, our Labrador?

We are now moving onto philosophically speculative ground, which inherently and by necessity is slippery. We just do not know final philosophical truths "for certain." But this must not stop us, not if we want to understand, even imperfectly, why we and animals matter.

To add to our exploration, I want to discuss briefly three philosophers boldly willing to go the extra mile beyond science into speculative philosophy in order to interpret why we, animals and nature matter—Aldo Leopold, Alfred North Whitehead and Hans Jonas.

Aldo Leopold.

The patron saint of modern conservation ethics, Aldo Leopold is famous for his land ethic and his defining the human good and bad in terms of our impact on the "beauty, stability and integrity" of the biotic community or the land (all biotic and abiotic elements of nature, including ourselves, that are implicated in life processes). In A Sand Country Almanac, Leopold explicitly embraces a Darwinian evolutionary and ecological scientific and philosophical world view; he exhorts us to become "plain members and citizens" of the land, rather than the conquerors of an alien or anti-human nature. The final piece of his philosophical revolution is the recognition of the genuine subjectivity and individual liveliness of animals—the she-wolf of "Thinking Like a Mountain," other top predators such as bears and cougars, as well as less imposing sandhill cranes, grebes, woodcock, chickadees, hunting dogs, brook trout, and more.

Leopold's philosophical revolution or conversion is perhaps most vividly expressed by his encounter with the wolf in "Thinking Like a Mountain."[4] Leopold had professionally been involved in game management in the Southwest, specifically the eradication of predators (wolves, bears, mountain lions) for the sake of increasing deer populations for hunters, if not paving the way for cattle ranging. While on a mountain trip, Leopold and his companions came upon a she-wolf crossing a river to join her cubs. Following the dictates of "wise-use" game management and the "trigger-itch" of young hunters, the group shot at the wolves, killing the she-wolf. As Leopold watched the "fierce green fire" dying in the wolf's eyes, he was taken up short. The mountain and the wolf knew something that Leopold did not, and what they knew put him to shame. Predators have an ultimate significance and central role to play in evolutionary, ecological and geological time and the ongoing well-being of ecosystemic nature and the humanly good life. Leopold had previously been thinking, feeling and acting in a wrong frame of reference. He had not taken a long-range, biotic, evolutionary, ecological perspective, and thus did not appreciate the roles that wolves and other large predators play in the overall health of specific ecosystems—keeping prey at healthy levels, preventing the overcropping of plant resources, helping the ecosystemic whole maintain a "dynamic balance or equilibrium." Henceforth, he knew better.

Two things are happening simultaneously here, both centrally important to philosophy and ethics. Leopold is radically converted to a biotic, evolutionary and ecological world view, but he also recognizes the genuine individuality and subjective life of animals—the "fierce green fire" of the wolf's eyes. Thus, Leopold could have a personal and "philosophical" relation with animals as well as wild flora that he encountered, precisely because he was convinced that he and they originated out of, and interacted within, evolutionary, ecological nature—that all life had a common origin and lively connection, a single historical home that includes all life and all lives. Leopold recognized a specific human difference or significance within nature—only humans can comprehend (though imperfectly) the evolutionary story and be

---

[4] Leopold, A. 1968. "Thinking Like a Mountain," <u>A Sand Country Almanac</u>. New York: Oxford University Press.

morally moved and motivated by its triumphs and tragedies. But this difference can only arise because we share varying evolved capacities of life with all other animal and animate life. All are "commonly connected" by the very fact of life, which variously expresses itself and its capacities in its individual instances. Animals importantly matter, but their mattering to us involves both their individual liveliness and our common historical context. Meeting animals on more equal terms in a common historical world is a characteristic theme in <u>A Sand Country Almanac</u>.

Alfred North Whitehead.

Earlier in the 20th century, Alfred North Whitehead made a similar philosophical and particularly telling move. In his classic essays, "Nature Lifeless" and "Nature Alive," Whitehead traces the historical rise and demise of modern Galilean-Cartesian-Newtonian science—a scientific world view of nature based dominantly on "superficial" common sense and sense perception (especially vision) and which assumes "nature at an instant" (no necessary and significant historical dimensions and interaction) and a "billiard-balls-in-motion" causal determinism and materialism.[5] Together these conceptions led by the end of the 19th century—with the new emphasis on electromagnetism, energetic dynamism, and field theory (the complex interconnections and interactions of all physical constituents of the world)—to what Whitehead claimed to be a "mystical chant over an unintelligible universe," a philosophical, if not scientific, dead end.

Whitehead was particularly interested in what modern scientific and philosophical positivism left unintelligible: the interconnections of life, mentality and nature, as well as fact and value, which are both primary derivatives of our everyday human experience. In a revolt against contemporary forms of materialism and idealism, Whitehead recoiled and philosophically dug deep into intimate bodily feelings. Here he found "nature alive," genuine episodes of natural liveliness, characterized by *absolute self-enjoyment* (an immediate, emotional apprehension or taking account of the antecedent world), *creative activity* (the individual process of apprehension), and *aim* (the aim at becoming a uniquely individual instance of life). Here he found the identity of the immediate self with the past self, with the organic body, and with the world abroad—the self in the world and the world in the self (mutual immanence). Here he found life, mentality and nature together— antecedent fact (achievement) and immediate value/valuing (emotional appreciation and lively response).

---

[5] Whitehead, A.N. 1968. <u>Modes of Thought</u>. New York: The Free Press.

Whitehead's insight is particularly helpful in our search for a relevant framework within which to articulate our proper relation as humans to the animal and natural world. Armed with his new interpretive conceptual scheme of "nature alive," Whitehead made intelligible what positivistic materialism left unexplained, if not incomprehensible: memory (the past functioning in the present), perception of the body and the external world, causation (efficient and final), why we arise where and when we do (the primacy of historical context, becoming, and natural, human and cultural evolution). Moreover, Whitehead rediscovered a significant, valuable and meaningful nature—instances of life (including animal and human individuals) interconnected and aiming individually and collectively at particular worldly achievements and intensities of experience. This "democracy" of historical and "community" life imbued Whitehead with a naturalistic, cosmological spirit (aesthetic and philosophical enjoyment) and appreciation of animal and animate life akin to Leopold's, though he was characteristically more interested in human intellectual and cultural history. The telling critique and grand synthesis of "Nature Lifeless" and "Nature Alive" are eminently worth re-reading if we want to comprehend how and why we and animals matter together.

Hans Jonas.

From a quite different mood and philosophical perspective, Hans Jonas also revolted against the hegemony of modern materialist science. He likewise found his way to a view of organic life that philosophically, morally and spiritually rehabilitated humans and nature and which served as a basis for envisioning important ethical responsibilities to the human and natural future.[6] As an active Jewish participant in the humanly fateful events of the Second World War, Jonas was particularly concerned with the specters of philosophical and moral nihilism, the denial of all moral truth, value and goodness. An intellectually naïve and unreflective scientific materialism (mechanistic determinism) denies the freedom and efficacy of mind and annihilates all subjective agency and achievements (realized values), human and other. It is a philosophical and ethical nihilism that silently haunts our contemporary world, undermining the legitimacy of moral, political and civic effort and authority. But such an uncompromising materialism, according to Jonas, leads to manifest absurdity and incoherence. Darwin's nature is rendered absurd by implanting in us humans the useless illusion of purpose, circumscribed freedom, and moral responsibility for no reason. Moreover, the philosophical materialists undercut their own rational

---

[6] Jonas, H. 1984. The Imperative of Responsibility. Chicago: University of Chicago Press. Also Donnelly, S. 1989. Hans Jonas, The Philosophy of Nature and the Ethics of Responsibility, *Social Research* 56:3: 635-657.

argument by denying all efficacy to mind and thought (their "epiphenomenalist" thesis). Why should we believe the materialists if they self-cancel their own rational authority? We are at a philosophical and spiritual midnight here.

In a philosophical move that parallels Whitehead's, Jonas denies that materialist science needs to turn philosophical and metaphysical, and pronounce upon reality according to its own partial and abstract conceptual scheme. This humanly characteristic blunder is what Whitehead terms the "fallacy of misplaced concreteness"—taking the abstract or partial aspects of things for concrete reality, the fullness of things. For Jonas, materialistic science should stick to its own limited guns of pursuing objective, empirically verifiable, "causal" knowledge— which is, importantly, not all that we know and understand and can experience about ourselves and the world, and especially the efficacy or reality of freedom and responsibility. Such intellectual circumspection or modesty gives Jonas the speculative philosophical opportunity to return to everyday experience, move (like Whitehead) beyond inherited modes of materialism and idealism, and systematically explore the "nature purposive" that is manifest in all organic, evolutionary life, animals and ourselves included.

Jonas understands us and animals to be psycho-physical individuals with common but varied forms of organic identity and living capacities.[7] All genuine instances of organic and especially animal life enjoy some form of self-feeling, selfhood, "needful freedom," and active commerce with the world. All have an important value and significance for themselves, if not also others. These fundamental features of organic life are engendered and necessitated by organisms' metabolic mode of existence, their active use of the "energetic" world in order to be their bodily selves and maintain their precarious mode of existence.

Metabolic and organic needs and capacities link humans and animals in one world of significant activity and achievement, with the primary goal being (organically) to be or not to be. This primal existential and purposive endeavor, "to be," shared alike by humans and animals, is the ultimate concrete value of reality, serving as the ground of value and the "good in itself," which needs no further justification, only our emphatic recognition. We humans have ultimate ethical responsibilities to uphold nature purposive, this intrinsically valuable good in itself—and especially purposive human nature, with its extraordinary moral, philosophical and spiritual dimensions. These fundamental moral responsibilities face the daunting challenges of burgeoning modern scientific and technological societies, which take little long-term heed of nature purposive.

---

[7] Donnelley, S. 1995. Bioethical Troubles: Animal Individuals and Human Organisms. *The Hastings Center Report* 25:7:21-29.

Final reflections. Here we have brief sketches of three different but important philosophical roads leading to a Rome at which we have yet to arrive: a full or at least adequate understanding of why we and animals philosophically and morally matter, and matter together. These Darwinian, naturalist, organicist perspectives put laboratory animal research and its supporters in an interesting moral situation. Neither Leopold, Whitehead or Jonas is against the use of animals for morally legitimate human purposes, including (presumably) scientific research. Indeed, a Darwinian evolutionary and ecological view of nature lends a powerful argument for legitimate uses of animals. Life must use itself in order to be and to realize its various values and goodness. Metabolic and organic rules govern our worldly realm. "Good" (benefit) and "evil" (harm) are intractably entwined.

Yet we and animals are also inextricably bound together physically, subjectively or experientially, and morally. Either both are morally significant, as claimed by these three philosophies of organism, or neither is finally significant, as implied by materialism and determinism as philosophies of reality. If we want to argue morally for the significance of science and lab research for the sake of human knowledge and practical benefits to humans and animals, then we must treat laboratory animals with genuine moral concern, care and respect— if we are to live morally coherent lives. We have no good reason to slip into cynical moral nihilism, fall into the fallacy of misplaced concreteness, and forget humans' and animals' common origin, togetherness, and significance in evolutionary, ecological nature. Too often this goal is absent in the human province—our labs— but it is much needed if we are fully to remember that we, animals, and nature really do matter. We need a robust and ever more adequate philosophy of organic life if we are to keep our moral heads straight and meet our responsibilities to the present and the future of humans and nature.

ENVIRONMENTAL VALUES IN PEACE AND WAR
Arthur H. Westing

Introduction

In order to examine current environmental values in peace and war, it seems necessary to begin by exploring societal dynamics with special reference first to social and then to environmental conditions during the second half of this century. It next becomes necessary to consider how one might determine societal values and thus their evolution over recent decades. The subsequent look at environmental values will be carried out within the context of overall societal values in both the civil and military sectors of society. All of this will lead up to the more important lessons that might be gleaned from the foregoing analysis. The present study builds upon prior work by the author.[1]

Societal dynamics

Societal dynamics can be divided into two spheres, the social and the environmental. The social dynamics of interest here are primarily those referring to human aggregations and how they interact. The environmental dynamics of interest here are primarily those referring to the global biosphere, its use and misuse.

Social dynamics: The matter of peace and war.

Human society has sorted itself into a huge array of social groups, large and small, more or less ephemeral, and only partially mutually exclusive. These many social groups derive their coherence and level of permanency from one or more affinities, among them especially: geographical, political, ethnic, linguistic and religious. The most clearly defined and delimited of these social groups are the sovereign states of the world, a status sanctified by the two Treaties of Westphalia of 1648 and variously reinforced ever since that time. The current number of generally recognized sovereign states is 192. Both within and overlapping these 192 coherent social groups there exist many hundreds of additional, less formal social groups enjoying varying levels of coherence, power, and permanency.

The often divergent interests and aims of these many formal and informal social groups inevitably lead to disputes between them. Fortunately for human society, the overwhelming majority of those inter-group disputes are resolved peaceably, that is, without resort to war. ("War" is defined as organized armed conflict with deadly and destructive intent.) However, throughout human existence a relatively

---

[1] Westing 1996; 1997; 1998.

modest fraction of those disputes have been settled by resorting to war. Nowadays, a few dozen wars of several to many years' duration are always in progress somewhere or other in the world, a frequency of occurrence that has fluctuated little during modern times, and certainly not much during this century. Thus merely since the end of World War II, the social groups of the world have engaged in hundreds of wars. Curiously, though, despite the increasing number of sovereign states in the world since that time—53 at the end of World War II as compared with 192 today—the number of international wars has been declining while the number of internal wars[2] has been rising.

In fact, today some 163 of the 192 sovereign states consider it necessary to devote a significant portion of their precious resources (material, financial and intellectual) to maintaining armed forces on a continuing basis.[3] The states—that is, their governments—do this in order to be able to deal with threats to their stability and tranquility, whether those be external or internal, accomplishing this through the counter-threat of war or else through war itself. Thus, for a great majority (85%) of all the sovereign states, and for virtually all of the major ones, it is really necessary to divide their society into two sectors: a military sector and a civil sector. Although the formal military sector of a state typically accounts for only a relatively small fraction of its society (say, 3% to 11%, depending on the measurement criteria employed[4]), it is an important fraction because it deals with what are widely deemed to be supreme national interests. Even more germane, in many states the military sector actually governs the civil sector.

To complicate matters further, in several dozen of the 192 states there exist internal social groups that are dissatisfied enough with their existing lot to establish and support irregular armed forces (e.g. insurgent forces, guerrilla forces) that must be considered for at least some purposes as another military sector distinct from the one composed of the regular armed forces.[5] Such irregular armed forces could be at war with the regular armed forces of their government or even with other internal irregular armed forces. Although these irregular groups come and go over the years in any one state, worldwide their total frequency in recent decades has varied little.

So to summarize the relevant social dynamics, most states and the social groups within them are "at peace" with each other (i.e. they interact non-violently) for most of the time, but nonetheless consider war a fully acceptable means of solving any seemingly intractable dispute.

---

[2] Non-international wars, internal wars, armed conflicts not of an international character.
[3] ACDA 1997, p. 36.
[4] ACDA 1997, Table 1.
[5] Sollenberg & Wallensteen 1998.

This approach to dispute resolution plays itself out within a human society which consists of a relatively large civil sector, a relatively small but important formal military sector, and an even smaller though also consequential informal military sector. These three sectors are in more or less intimate contact with each other, interacting with one another in various constructive or destructive ways. And it must be remembered that for any one state, the sectors present—whether two or three (or even more)—may well consist largely of individuals drawn from the same societal pool.

Environmental dynamics: The biosphere, its use and misuse. Recent decades have been punctuated by a number of environmental catastrophes of human origin that have certainly helped to focus worldwide attention on the human environment. I shall name but the few most salient ones: the Second Indochina War (Vietnam Conflict) of 1961-1975, the Seveso dioxin release of 1976, the *Amoco Cadiz* oil spill of 1978, the Bhopal methyl isocyanate release of 1984, the Chernobyl radioactive nuclide (iodone-131, cesium-137, strontium-90, etc.) release of 1986, the *Exxon Valdez* oil spill of 1989, and the Persian Gulf War of 1991. As important as these calamitous events have been, they should not distract us from the big picture of changes in our routine environmental use and misuse over time.[6]

The global biosphere has, of course, been utilized by human society since the beginning of our existence, both directly and indirectly.[7] However, it has not been over-utilized until quite recently in our lengthy history. In 49 BC, a soldier named Julius crossed the Rubicon, having thereby made a conscious and irrevocable decision to conquer Rome or perish in the process. Almost exactly 2000 years later, our human society crossed its Rubicon. We did so by continuing beyond sustainability our somewhat more grandiose attempt to conquer not merely Rome, but the entire earth. Human society made the decision to cross the Rubicon rather less thoughtfully, it seems, than did Julius, this despite considerably greater access to pertinent information. We can only hope that it was not equally irrevocable.

Human actions—driven by our ever growing numbers, abilities, needs, and desires—did not exceed sustainability until roughly four decades ago. It was at about that time that the major renewable resources of the earth upon which our very survival depends (agricultural soils, forest products, rangeland grasses, and ocean fishes) began to be used beyond their sustained ability to renew themselves. It was also at about that time that the major waste depositories of the earth (the land, the waters, and especially the atmosphere) began to be utilized for the disposal of our routine solid, liquid, and gaseous wastes beyond their

---

[6] Polunun 1998; Westing 1980.
[7] Costanza et al. 1997; Daily 1997.

sustained ability to dissipate and decompose them. It was also at about that time that we began to encroach upon wild and semi-wild nature at levels that diminish biodiversity at flagrantly accelerated rates.[8] And it was only a few years before that time that we began to develop weapons that have the true potential to cause widespread, long-term severe damage to the environment.

So to summarize the relevant environmental dynamics, utilization of the earth's natural resources—both for the goods being extracted and for the wastes being injected—has been steadily increasing throughout at least this past half century, crossing the threshold of sustainability perhaps 40 or so years ago, nonetheless to this day continuing its trajectory. But as we shall soon see, these destructive habits have not gone entirely unnoticed by the world.

Determining societal values

Being centrally concerned here with environmental values, and perforce also with the overall societal values in which they are embedded, we face the crucial question of how to determine what those values are. How do we reach into the psyches of the five billion and more people who now comprise human society, and then aggregate those findings? I have decided to approach this dilemma primarily though an examination of two lines of evidence, the resolutions of the United Nations General Assembly on the one hand, and the commitments entered into via multilateral treaties on the other.

The United Nations General Assembly has been gathering at least annually since its birth at the end of World War II, doing so in both regular (annual) and special (occasional) intergovernmental sessions and conferences. Each time these Assemblies bring together the representatives of most of the sovereign states of the world, beginning with 55 at the first Assembly in 1946, and progressively building up to the 185 which have come together at the 53rd Assembly currently in progress (Fall 1998). At each of these many sessions the national delegates, acting on behalf of their respective governments, ultimately accept many hundreds of resolutions. It must be stressed that these many General Assembly resolutions are only hortatory and non-binding in nature, and are thus often referred to as "soft law." Nonetheless, those brought before the Assembly for consideration are seriously debated by the delegates, often at great length and sometimes accompanied by intense lobbying, before being rejected or accepted, whether by acclamation or recorded vote.

I therefore consider the expressions of societal values in the General Assembly resolutions (when considered in conjunction with the associated voting records) to be a valuable source of relevant, tangible

---

[8] Baillie & Groombridge 1996; Groombridge 1992; Walter & Gillett 1998.

information. Of course, this distillation—this aggregation of what is contained in five billion psyches worldwide—presupposes that the positions of the 185 national delegations actually reflect not only the wishes of their respective governments, but that those governmental positions in turn actually represent the values of their respective citizenries. In that regard it must be noted that perhaps 40% of the states in the world are governed democratically and another 32% partially so, thus leaving 28% under despotic rule.[9] But this record is not as bad for present purposes as it might at first seem, because all the delegates to the Assembly attempt, on behalf of their respective governments, to exude an aura of respectability, giving them at least the opportunity to express national feelings in voting on non-binding resolutions.

Regarding international treaties, sovereign states have been entering into binding agreements with each other—whether bilateral, regional, or worldwide—for centuries now. During its relatively brief tenure, the League of Nations registered a total of approximately 5000 international treaties, both bilateral and multilateral, between World Wars I and II. The United Nations took over this task following World War II and has to date (as of Fall 1998) registered close to another 35,000 of them. In keeping with the time frame adopted, I am here considering primarily multilateral treaties entered into during the past half-century, restricting myself to the major (that is, widely adopted) ones.

A multilateral treaty considered for acceptance by a state undergoes a review process that is even more elaborate and lengthy than the one for a United Nations General Assembly resolution. It is a major step for a sovereign state to become party to a treaty because in doing so it gives up a bit of its precious and tenaciously held freedom of action— that is, a bit of its sovereignty. A nation therefore submits to a treaty only when that instrument reflects its position (or at least its public position) on the matter at hand and when the gains it perceives to be thereby achieved outweigh the concomitant loss in sovereignty. In short, multilateral treaties reflect societal attitudes in a manner similar to that of the resolutions discussed earlier, as an indicator sharing both their strengths and weaknesses, but complementing them to the extent that they differ in the intensity of consideration to which they are subjected prior to adoption or rejection.

Evolving societal values

The expansion of social concerns. The experience of World War II clearly helped crystallize international concern for social values. Two striking articulations of such social concern were the 1948 Convention

---

[9] Karatnycky 1998.

on the Prevention and Punishment of the Crime of Genocide[10] (to which
the United States is a relatively recent party [1988]) and the 1948 United
Nations Universal Declaration of Human Rights[11] (with the United States
among the original endorsers). Through the Genocide Convention, the
states party (now about two-thirds of the possible total) formally
confirmed that destruction of a national, ethnic, racial or religious group
constitutes a punishable crime. The soon to follow Human Rights
Declaration proclaimed that all humans are born free and equal in dignity
and rights, that everyone automatically possesses an array of rights,
important among them the right to life, liberty and the security of person;
the right to take part in the government of their country; the right to a
standard of living adequate for health and well-being; and the right to an
education. Although this lofty Declaration was, of course, only
aspirational in nature, most of its proclaimed human rights were
subsequently formally embraced by more than 70% of our 192 sovereign
states by virtue of their becoming party to two complementary (and
partly overlapping) treaties: the 1966 International Convenant on
Economic, Social and Cultural Rights[12] (to which the United States is not
a party), and the 1966 International Covenant on Civil and Political
Rights[13] (to which the United States is a recent party [1992]).

Indeed, a considerable number of social values were enunciated
by the international community in the several decades following World
War II, to be expressed in growing numbers of General Assembly
resolutions and to be formalized in a host of newly formulated
multilateral treaties. The treaties in question can be grouped into three
distinct categories of social concern, those regarding (a) human rights
issues; (b) wartime humanitarian issues; and (c) arms control and
disarmament issues. By way of example for each of these three
categories, one might point especially to: (a) the 1951 Convention
Relating to the Status of Refugees plus its 1967 Protocol; (b) the four
1949 Geneva Conventions Relative to Humanitarian Conduct in Times
of War plus their two 1977 Protocols; and (c) the 1972 Convention on
the Prohibition of Bacteriological and Toxin Weapons.[14]

This represented a veritable flood of interest during the first
quarter-century following World War II. It is remarkable that the
international community, striving to express and codify societal values,
focused exclusively on *social* concerns. That is to say, those expressions
were unencumbered by environmental concerns—neither for the

---

[10] UNTS 1021.
[11] UNGA Res 217 (III)A.
[12] UNTS 14531.
[13] UNTS 14668.
[14] UNTS 2545, UNTS 8791; UNTS 970-973, UNTS 17512 & 17513; UNTS
14860.

environment per se nor even for the environment as the basis for realizing the expressed social concerns. It is additionally noteworthy that the humanitarian and related values growing out of World War II all express in one way or another a concern for making warfare more humane, not for eliminating it. The widespread post-World War I concern to outlaw interstate warfare did, in fact, lead to the 1928 Treaty Providing for the Renunciation of War as an Instrument of National Policy[15] (adopted by virtually all states existing at the time, including the United States). But that utopian social value, embraced in the euphoric aftermath of the war to end all wars, soon shriveled on the vine to die a quiet death.

The expansion of environmental concerns. It took almost three decades following World War II—and roughly two decades following the crossing of our Rubicon—before the international community finally began to take serious notice of our increasingly unsustainable patterns of biospheric exploitation. The first major international expression of those newly emerging environmental values was via the 1972 United Nations Conference on the Human Environment, held in Stockholm. Widely endorsed, the Declaration of that Conference[16] proclaimed a significant amplification and extension of the 1948 Universal Declaration of Human Rights as its most fundamental "common conviction." The articulation of this conviction was complex and included principles of rights as well as responsibilities. It stated that humans have the fundamental right to freedom, equality, and adequate conditions of life in an environment that permits dignity and well-being; and also that human beings bear a solemn responsibility to protect and improve the environment for their present and future generations. Other important principles of the Declaration included the conviction that human beings have a special responsibility to safeguard and wisely manage the heritage of wildlife and its habitat; that economic and social development are essential for ensuring a favorable living and working environment; that education in environmental matters is essential in order to broaden the basis for an enlightened opinion and responsible conduct; and, finally, that humans and their environment must be spared the effects of nuclear weapons and all other means of mass destruction. In fact, this was a period of intense worldwide environmental awakening, with the environmental values enunciated in the 1972 Declaration reinforced by a number of specialized treaties, central among them the 1971 Convention on Wetlands of International Importance, the 1972 Convention on the World Cultural and Natural Heritage, and the 1973 Convention on International Trade in

---

[15] LNTS 2137.
[16] UNGA Doc A/CONF.48/14/Rev.1.

Endangered Species of Wild Flora and Fauna (with the United States among the many parties to all three of these instruments).[17]

The environmental values expressed in the early 1970s all derived from anthropocentric concerns. It was not until a decade later that the biosphere entered into the collective human psyche on ecocentric terms, albeit only in small part. This rather dramatic progress was demonstrated most prominently though the 1982 United Nations World Charter for Nature,[18] a hortatory proclamation that had but one country in the world—the United States—voting in opposition to it. The fundamental principle expressed by the 1982 Charter was that nature shall be respected and that its essential processes shall not be impaired. Moreover, it boldly proclaimed that all human conduct affecting nature was to be guided and judged by the principles of conservation it contained; and in a remarkable departure for the United Nations, the final principle of this Charter proclaimed that (in addition to each state) each person has a duty to act in accordance with its provisions. The remaining principles were to a considerable extent repetitive of the 1972 Declaration, if a bit more incisively expressed. For example, in the planning and implementation of social and economic development activities, due account was to be taken of the fact that conservation of nature is an integral part of those activities. Also, nature had to be secured against degradation caused by warfare or other hostile activities; and (peacetime) military activities damaging to nature had to be avoided. Among the explicit reasons given by the United States for its opposition to the 1982 Charter were: (a) that the Charter lacked clarity and precision; (b) that it was too prescriptive (that all of its "shalls" ought to have been "shoulds"); and (c) that obligations should not have been assigned to individuals, but only to nations.[19]

It is clear that the pro-environmental momentum gathered in the early 1970s actually peaked in the early 1980s, having over-reached itself in the 1982 Charter. Thus, the next major expression of international environmental values came a decade later in the form of a much heralded treaty, the 1992 Convention on Biological Diversity (to which over 90% of countries in the world are now a party, although not including the United States).[20] This treaty, one of the few important outcomes of the 1992 United Nations Conference on Environment and Development (held in Rio de Janeiro), represents a substantial retrogression from the environmental values that had been espoused in the early 1980s. The Rio Conference Declaration itself,[21] although again

---

[17] UNTS 14583; UNTS 15511; UNTS 14537.
[18] UNGA Res 37/7.
[19] Westing 1987.
[20] UNTS 30619.
[21] UNGA A/CONF.151/5/Rev.1.

largely repetitive of the 1972 and 1982 declarations, no longer espoused ecocentric justifications. Indeed, it proclaimed as its fundamental principle that humans must be at the center of environmental concerns. Even more definitively, the Biological Diversity Convention (beyond its valuable aspects) had three major shortcomings.[22] First, despite some lip service to the contrary, it effectively rejects the concept of sustainable development by virtue of assigning overriding priority to economic development.[23] Second, it actually repudiates the principle that global biodiversity is a common heritage of humankind, thereby denying the responsibility of the community of nations to act in concert to protect that heritage.[24] Third, through its silence on the matter, it denies the responsibility of the community of nations to act in concert to protect biodiversity in areas beyond national jurisdiction (the extra-territorial domains, in this instance especially the high seas).[25] It is of considerable interest to point out that none of the reasons given by the United States for rejecting the Biological Diversity Convention coincide with the three flaws just enumerated. Rather, the declared dissatisfactions of the United States deal particularly with the treatment of intellectual property rights, finances, technology transfer, and biotechnology.

My measures of change in environmental (and other societal) values could be usefully supplemented or verified by various other approaches—public opinion polls, growth in particular public agencies and in nongovernmental organizations, voting records for "green" candidates, and domestic legislative histories. However, the necessary public opinion polls are in essence available only for the two dozen or so industrialized—and, for the most part, democratically governed—countries which together account for less than 20% of the global population. And worldwide data on the other possible measures would be quite tedious to assemble. Two unrelated worldwide environmental opinion polls do exist, both fairly recent. The first of these surveyed 16 countries on five continents during 1988-1989 relying, however, on rather small samples;[26] and the second surveyed 24 countries on the same five continents during 1992 relying on somewhat larger samples.[27] In support of my findings, both reveal strong public concern for the environment, *irrespective* of the level of wealth or industrialization of the country.

Neither of these two public opinion polls provides clues to changes in attitude over time. On the other hand, we can get an

---

[22] Guruswamy 1998.

[23] Cf. Article 20.4; see also Preamble para. 19.

[24] Cf. Article 3; see also Preamble para. 3 & 4.

[25] Cf. Articles 3 & 4.b.

[26] Bloom 1995; Tolba & El-Kholy 1992, pp. 674-675.

[27] Bloom 1995; Dunlap et al. 1993.

indication of globally evolving environmental values from the establishment of relevant governmental agencies.[28] Indeed, also supportive of my findings, in 1972 (at the time of the Stockholm Conference on the Human Environment) fewer than ten countries had environmental ministries or comparable national bodies. Sweden's was the first, founded in 1969, very soon to be followed by the United States' Environmental Protection Agency (EPA), founded in 1970, the United Kingdom, Canada, Japan, France, Norway, New Zealand, and Australia. But by 1992 (at the time of the Rio Conference on Environment and Development), more than 50 diverse countries had them.

Moreover, some time series are available specifically for the United States that also suggest a progressive development in our national environmental values. For example, seven comparable polls taken during the period 1981-1990 reveal a fairly steady increase in that fraction of the public which is greatly concerned with the state of the environment, rising during that decade from 45% to 75%.[29] And membership between 1970 and 1990 increased five-fold in both the National Audubon Society (with a steady rise from 105,000 to 515,000) and the Sierra Club (with a progressive rise from 114,000 to 566,000).[30]

Environmental concerns in relation to the military sector in peacetime. Thus far in my exposition of evolving societal values I have made no distinction between the civil and military sectors of society. I see these two sectors as having been cut largely from the same cloth, thus largely sharing societal values, whether social or environmental. Indeed, the evolution since World War II of environmental concerns within the military sector appears to have paralleled that within the civil sector, as have responses to such concerns. Nonetheless, it is instructive to single out environmental values in the military sector in support of this contention, doing so first regarding peacetime and then regarding wartime.

Well over a hundred separate sovereign states have made hostile use of their regular armed forces at one time or another since World War II.[31] Nonetheless, most of the world's armed forces are not engaged in hostile activities most of the time. Rather, they are busy in such routine peacetime functions as training, garrison duty, patrolling, weapons testing, and, of course, serving as a threat to potential enemies. Moreover, as already noted, in some instances they are at the same time running their governments, whether overtly or covertly. However, even in the democratic states, the formal military sector is often legally

---

[28] Tolba & El-Kholy 1992, Ch. 22.
[29] Tolba & El-Kholy 1992, p. 672.
[30] Id., p. 680.
[31] Tillema 1989.

beyond the reach of the domestic laws and regulations that control environmentally relevant actions. By way of example, the armed forces of Germany, Serbia/Montenegro, Switzerland, and the United Kingdom are, in whole or in part, exempt from their states' domestic environmental protection legislation.[32]

On the other hand, national environmental protection legislation applies equally to the military and civil sectors, at least domestically during peacetime, in at least 19 states—Bangladesh, Croatia, Denmark, Finland, India, Indonesia, Iran, Malaysia, the Maldives, the Netherlands, Norway, Pakistan, Poland, South Africa, Sri Lanka, Sweden, Thailand, Vietnam, and the United States—a goodly mix of democratic and autocratic states.[33] Moreover, the North Atlantic Treaty Organization (NATO) has recently developed a set of quite detailed environmental guidelines for the armed forces of its 16 member states during peacetime, going on to suggest that these would be appropriate for any state to adopt.[34] The NATO guidelines promote environmental responsibility and in essence urge that, within limits, the military sector of a state should comply with the environmental rules established for its civil sector. Indeed, through its own sound environmental practices, the military sector should, as NATO would have it, be serving as an example to the rest of the country.

Even in the absence of military/civil parity before domestic law, it is clear that environmental concerns are beginning to pervade the regular armed forces of the world. The defense ministries of at least 11 disparate states have in recent years established permanent environmental departments and programs: Bulgaria, Croatia, the Czech Republic, Denmark, Germany, Hungary, Pakistan, Sweden, the United Kingdom, the United States, and Vietnam.[35] The United States appears to have done this more thoroughly and elaborately than any other state, with one high Pentagon official proudly referring to the United States armed forces as now being "lean, mean and green." Moreover, for better or worse, 13 or more states assign to their armed forces the enforcement of their domestic environmental protection laws: Bangladesh, Bhutan, Cambodia, India, Indonesia, Laos, Malaysia, the Maldives, Myanmar, Nepal, the Philippines, Sri Lanka, and Thailand.[36]

The formal military sector of a state is likely to be acting in an environmentally responsible way, because in so doing it is acting in its own self interest. The individuals who make up the military sector have just as much at stake as those who make up the civil sector in

---

[32] Westing 1998.
[33] Id.
[34] NATO 1996.
[35] Westing 1998.
[36] Id.

maintaining an environment conducive to national survival and well-being. Moreover, in the more or less democratic countries of the world the military sector is always in competition for its share of the national budget. It must thus strive to establish and maintain as good a public image as possible, not only as defender of the nation from any hostile forces and supporter of its foreign policies, but also as an exemplary member of domestic society. And recall that the military sector (at least the formal military sector) is composed of individuals who have gone through their country's educational system, are exposed to their country's news media, send their children to their country's schools, belong to all sorts of conservation organizations, have such outdoor interests as bird watching, and—overall—consider themselves to be responsible, contributing members of their society.

Environmental concerns in relation to the military sector in wartime. It is easy to suggest that the true environmental values embraced by the military sector of a society will not be revealed during peacetime, but rather through the actions of its armed forces when engaged in warfare. Any such examination requires that we distinguish between interstate (international) and intrastate (domestic) warfare. The codification of environmental values distinguishes between these two forms, and the differences are considerable.[37]

As to interstate warfare, about 80% of all countries (though not the United States) have adopted 1977 Protocol I on International Armed Conflicts,[38] which commits them, in warfare between themselves, to take care to protect the natural environment of their adversary against widespread, long-lasting, and severe damage. On the other hand, it must be noted that only about one-third of the states party to this treaty have demonstrated a true (albeit sovereignty-eroding) commitment to this constraint by virtue of opening themselves to a compulsory verification procedure in the event of an alleged breach.[39] Moreover, self-inflicted wartime destruction of the environment is explicitly permitted to states party to this treaty. 1977 Protocol I is an instrument of central importance for our concerns because it expanded international humanitarian law to embrace environmental values. The United States' opposition to this Protocol derives from several issues.[40] The most important of its announced problems with this landmark treaty was the inclusion of wartime protection of the natural environment; that principle was said to be presented in too broad and ambiguous a fashion for the United States to accept. The second major objection was the constraint

---

[37] Schmitt 1997-1998; Westing 1997.
[38] UNTS 17512.
[39] Cf. Article 90.
[40] Roberts 1998.

against attacking works and installations that had the potential for releasing so-called dangerous forces (e.g. radioactive nuclides).

Despite the United States' rejection of 1977 Protocol I, the United States Navy and Marine Corps recently saw fit to issue the following directive on "Environmental Considerations" to all their commanders—apparently the first inclusion of the sort into the military manual of any armed force in the world:

> It is not unlawful to cause collateral damage to the natural environment during an attack on a legitimate military objective. However, the commander has an affirmative obligation to avoid unnecessary damage to the environment to the extent that it is practicable to do so consistent with mission accomplishment. To that end, and as far as military requirements permit, methods or means of warfare should be employed with due regard to the protection and preservation of the natural environment. Destruction of the natural environment not necessitated by mission accomplishment and carried out wantonly is prohibited. Therefore, a commander should consider the environmental damage which will result from an attack on a legitimate objective as one of the factors during targeting analysis.[41]

Turning to intrastate warfare, 1977 Protocol II on Non-international Armed Conflicts[42] contains no constraint on environmental damage comparable to the one just noted for an adversary's territory in interstate warfare. In fact, this entire treaty is modest indeed, in its various prescriptions and proscriptions, the drafters having presumably recognized the potential for rejection on the basis of undermining the so widely and dearly held concept of national sovereignty. Perhaps equally important was the fear that acceptance of a more strongly worded instrument might tend to legitimize and even encourage insurgencies,[43] one of the major expressed concerns of the United States. The real problem here is that all gradations of organized intrastate dispute can occur, ranging from non-violent protest to internal disturbances, tensions, or riots (thus a domestic issue to be dealt with by civil police)—and on to outright insurgency warfare. Especially in the case of an intrastate war in which the irregular armed forces are attempting to secede, those irregular armed forces do, in fact, sometimes announce to the world at large that they intend to abide by international humanitarian law, doing so to gain respect and legitimacy both at home and abroad.

---

[41] US Navy et al. 1995, para. 8.1.3.

[42] UNTS 17513.

[43] Lopez 1994.

150

Although not in force, the 1998 Statute of the International Criminal Court[44] lists among its catalog of specific war crimes applicable to international wars an attack in the knowledge that it will cause widespread, long-lasting, and severe damage to the natural environment that would be excessive in relation to the anticipated military advantage.[45] On the other hand, no comparable war crime is listed for non-international wars. The United States has stated its opposition to this instrument in essence as being too destructive of our national sovereignty.

Conclusion

At least three important concerns emerge from the foregoing discussion that in turn suggest components of an agenda for the future.

A **first** concern is the United States' poor record in supporting international environmental obligations, despite the strong environmental values that are (as have been revealed by sequential public opinion polls) so widely and ever more strongly shared in our country. Our hesitancy to join most of our allies and the numerous other nations in the world in so many of the key formal expressions of environmental concern and commitment seems to derive from some combination of at least five factors:

(a) a tenaciously held commitment to national sovereignty (with concomitant aversions to world government and to interference with affairs considered internal);

(b) a feeling that our position of global superiority—economic strength, military power, wealth of natural resources, adequate domestic legislation and programs—largely obviates the need for cooperative efforts on our part;

(c) an historical tendency toward isolationism (possibly including some xenophobic overtones) that does not grasp the full implications of the global biosphere;

(d) a strongly held notion that the extra-territorial domains of the world are to be exploited on a first-come/first-take basis; and

(e) a tendency to assume that environmental problems affecting us can be overcome by technological advances ("fixes").

As a **second** concern, we must somehow mitigate the environmental impact of the now rather frequent instances of irregular armed forces engaging in intrastate warfare, a frequency of incidence that has every likelihood of continuing into the future. In the almost complete absence of applicable

---

[44] UNGA Doc. A/CONF.183/9.
[45] Article 8.2.b.iv.

international law, and because insurgent forces are by their nature beyond the reach of domestic environmental (or other) law, we must seek other avenues to deal with this problem. Thus, if for no other reason (and there are, of course, others), it becomes important for us to foster environmental values everywhere in the world. Such nurturing of environmental values must be for the entire population of a country and it must begin at an early age. This is important so that those who might later in life join insurgency movements already have established in them a firm tendency to respect the environment. We should therefore encourage, offer expertise to, and financially support environmental education on a worldwide basis, both formal and informal and for all age levels. One way to do this would be via the joint UNESCO/UNEP[46] International Environmental Education Program. Such educational efforts should, of course, be aimed at both the civil and military sectors—and in each of those sectors addressing both peacetime and wartime issues.

As a **third** concern, it is crucial for the global biosphere—and thus also for the future well-being of humankind—that the democratic, non-corrupt, and developed nations of the world strive to spread those social attributes throughout the world. I base this on my finding that as a nation improves in measures that demonstrate those attributes the likelihood of that nation accepting humanitarian constraints having environmental ramifications increases dramatically.[47] Some additional linkages emphasize the relevance of this third concern: others have found that international wars in which democratic nations engage are apt to lead to fewer fatalities than those of non-democratic nations; that democratic nations are unlikely to go to war with each other; and that democratic nations seem able to avoid conditions of domestic famine.[48]

Thus, it is only when the positive social and environmental values already prevalent among the peoples of the world can be tangibly expressed to their full extent that we will be able to entertain any hope for a satisfactory future, either for ourselves or for the other living things on earth. And the achievement of such a future would be enormously facilitated if the United States, the trendsetter for so much of the world, would (a) set a socially and environmentally constructive example to emulate in the sphere of international cooperation, and (b) provide relevant assistance to the many less advantaged countries and peoples of the world.

---

[46] United Nations Educational, Scientific and Cultural Organization/United Nations Environmental Program.

[47] Westing 1997. The measures I relied on there were level of democracy, degree of integrity, national wealth, industrialization, and human development—the last measured as an amalgamation of life expectancy at birth, adult literacy rate, school enrollment, and gross domestic product (per capita).

[48] Rummel 1995; Weart 1998; Sen 1996.

152

References

ACDA 1997. World Military Expenditures and Arms Transfers 1996. 25th ed. Washington: US Arms Control & Disarmament Agency.

Baillie, J. & Groombridge, B. Eds. 1996. 1996 IUCN Red List of Threatened Animals. 5th ed. Gland, Switzerland: World Conservation Union (IUCN).

Bloom, D.E. 1995. International public opinion on the environment. *Science*, Washington, 269:354-358.

Costanza, R. et al. 1997. Value of the world's ecosystem services and natural capital. *Science*, Washington, 387:253-260.

Daily, G.C., Ed. 1997. Nature's Services: Societal Dependence on Natural Ecosystems. Washington: Island Press.

Dunlap, R.E., Gallup, G.H. Jr. & Gallup, A.M. 1993. Of global concern: Results of the health of the planet survey. *Environment*, Washington, 35(9):7-15, 33-39.

Groomsbridge, B. , Ed. 1992. Global Biodiversity: Status of the Earth's Living Resources. London: Chapman & Hall.

Guruswamy, L.D. 1998. Convention on Biological Diversity: A Polemic. In Guruswamy, L.D. & McNeely, J.A., Eds. Protection of Global Diversity: Converging Strategies. Durham, NC: Duke University Press. 351-359.

Karatnycky, A., Ed. 1998 Freedom in the World: The Annual Survey of Political Rights & Civil Liberties 1997-1998. New Brunswick, NJ: Transaction Publishers.

Lopez, L. 1994. Uncivil wars: The challenge of applying international humanitarian law to internal armed conflicts. *New York University Law Review*, New York, 69:916-962.

NATO 1996. Environmental Guidelines for the Military Sector. Brussels: North Atlantic Treaty Organization, Committee on the Challenges of Modern Society.

Polunin, N., Ed. 1998. Population and Global Security. Cambridge: Cambridge University Press.

Roberts, A. 1998. The laws of war and environmental damage. Washington: Environmental Law Institute, in press.

Rummel, R.J. 1995. Democracies ARE less warlike than other regimes. *European Journal of International Relations*, London, 1:457-497.

Schmitt, M.N. 1997-1998. Green war: An assessment of the environmental law of international armed conflict. *Yale Journal of International Law*, New Haven, 22:1-109.

Sen, A. 1996. Freedom favors development. *New Perspectives Quarterly*, Los Angeles, 13(4):23-27.

Sollenberg, M. & Wallensteen, P. 1998. Major armed conflicts. *SIPRI Yearbook*, Oxford, 1998:17-30.

Tillema, H.K. 1989. Foreign overt military interventions in the nuclear age. *Journal of Peace Research*, Oslo, 26:179-196, 419-420.

Tolba, M.K. & El-Kholy, O.A., Eds. 1992. The World Environment 1972-1992: Two Decades of Challenge. London: Chapman & Hall.

US Navy, Marine Corps, & Coast Guard 1995. Commander's Handbook on the Law of Naval Operations. 3rd ed. Washington: US Department of the Navy, Office of the Chief of Naval Operations, Publication No. NWP 1-14M (formerly NWP 9 (Rev.A))/FMFM 1-10/COMDTPUB P5800.7.

Walter, K.S. & Gillett, H.J., Eds. 1998. 1997 IUCN Red List of Threatened Plants. 1st ed. Gland, Switzerland: World Conservation Union (IUCN).

Weart, S.R. 1998. Never at War: Why Democracies Will Not Fight One Another. New Haven: Yale University Press.

Westing, A.H. 1980. Warfare in a Fragile World: Military Impact on the Human Environment. London: Taylor & Francis.

Westing, A.H. 1987. World Charter for Nature. *Environmental Conservation*, Cambridge, UK, 14:187-188.

Westing, A.H. 1996. Core values for sustainable development. *Environmental Conservation*, Cambridge, UK, 23:218-225.

Westing, A.H. 1997. Environmental protection form wartime damage: The role of international law. In Gleditsch, N.P., Ed. Conflict and the Environment. Dordrecht: Kluwer Academic Publishers. 535-553.

Westing, A.H. 1998. In furtherance of environmental guidelines for armed forces during peace and war. Washington: Environmental Law Institute, in press.

ROE V. WADE AS A COUNTER-REVOLUTIONARY MANIFESTO: A RETROSPECTIVE VIEW

Robert A. Burt

The abortion dispute is commonly understood as a diametrically opposed conflict between proponents of different world-views. The pro-choice forces espouse radical individualism and moral relativism while the pro-life forces embrace radical communitarianism and moral absolutism—so goes the conventional depiction of the "abortion wars." There is, however, at least one fundamental belief shared by both sides in this warfare—a belief that contemporary America is not a reliably nurturant society where vulnerable people can trust that their neediness will be sympathetically met. The two sides disagree about whether America was ever reliably nurturant. The pro-life forces typically maintain that there was such a golden age, and one path to finding our way back there is to honor the nurturance needs of vulnerable fetuses. Among the pro-choice forces many feminists, in particular, assert that America was never reliably nurturant, that patriarchal authority disguised itself in a benevolent cloak but was always abusive and exploitative toward its subordinates. Self-reliant individualism—including the right of every woman to decide her own needs regarding abortion—is the answer endorsed by these pro-choice forces as the way toward establishing a new Jerusalem, an unprecedented America where vulnerable people are respected rather than abused.

These prescriptions are thus radically different, but the diagnosis of the underlying ailment—the absence of reliable social nurturance—is the same. It is not surprising that abortion should be one of the central subjects where this absence, this diagnosed ailment, is widely and intensely felt. For the pro-life forces, the conspiracy of forces arrayed against the utterly vulnerable fetus is the entire catalogue of traditional caretakers: mothers, doctors and (after *Roe v. Wade*) judges. The betrayal of trust could not be more complete. For the pro-choice forces, the pregnant woman occupies this vulnerable position—forced to nurture the fetus and the unwanted child while her own needs are unattended.

There is a sad irony in the public clash between the pro-life and pro-choice forces. The commonly shared conviction that America is not reliably nurturant toward vulnerable persons is not simply obscured by the adamant disagreements about the nature of society's "ailment," its etiology and the prescriptions for its cure. Rather, the noisy disagreements mask any acknowledgement of the extreme difficulties, perhaps even the impossibility, of changing this widespread belief. If the rival positions of these two forces are understood as answers offered to the problem, it is apparent that both sides are implicitly asserting that the problem is not only desperately urgent, but also utterly insoluble. The answers offered by each side, that is, seem calculated to confirm widespread beliefs about the unreliability of social caretaking.

The pro-choice position of reliance on individual choice and self-control is an attractive principle in the normative tradition of liberal individualism, but it offers little comfort as an answer to the underlying question, "who will protect me when I am too vulnerable and too needy to take care of myself?" This, however, is the basic problem posed by deep mistrust toward traditionally honored caretakers.[1] The pro-life forces appear to speak more directly to this basic problem, but they do so by choosing sides in an obviously arbitrary and unsatisfying way. The fetus is clearly vulnerable, but so is the pregnant woman who, for whatever reason, does not want this fetus and in some cases feels victimized by its caretaking demands. The pro-life forces try to refute this equal vulnerability by portraying the pregnant woman as typically selfish ("she wants an abortion just to look good in a swimsuit") and voluntary ("she chose to have unprotected sexual relations and should now accept the consequent caretaking responsibility"). But the refutation is not adequately convincing because the pro-life insistence on exclusive attention to the neediness of the fetus necessarily requires forfeiture of all possible sympathy for the neediness of the unwillingly pregnant woman. It is not logically or emotionally clear that the woman's neediness should necessarily take precedence over the fetus's, though the pro-choice forces try to establish this preferred status by insisting that the fetus is merely biological protoplasm while the woman alone is a "person." It is equally unclear, however, that the needy fetus deserves more protection than the needy woman. There is *a logically and emotionally insoluble conflict* here which the tub-thumping public clamor of the pro-choice and pro-life forces cannot effectively conceal. Their shared effort at noisy concealment is more a confirming denial than a convincing refutation of this essential insolubility.

The Supreme Court's decision in *Roe v. Wade*[2] took place against the background of this social dispute. The Justices implicitly understood that they too were offering an answer to a widespread social complaint that American society was not a reliably nurturant place. Close attention to the details of the Justices' opinions in *Roe* makes clear that they saw themselves as participants in the debate, but that—unlike the most adamant antagonists in the pro-choice and pro-life camps—the Justices did not accept the premise that this complaint was true. Given their social status as a high embodiment of traditional caretaking, it would have been quite surprising if the Justices had endorsed such a self-

---

[1] See Robert A. Burt, "The Suppressed Legacy of Nuremberg," 26 Hastings Center Report 30 (1996).
[2] 410 U. S. 113 (1973).

impeaching critique. It is apparent, however, that the Justices took it upon themselves to answer this critique—apparent not only from what they said in their opinions but even more fundamentally from the fact that seven members of the Court took the extraordinary (and quite unexpected) step of constitutionally invalidating most states' abortion laws. In what they said and what they did the Justices clearly announced their concurrence with the complaint that existing caretaking arrangements were inappropriate regarding the symbolically charged and practically important abortion issue. The Justices themselves claimed that they could point the way toward a more reliable future.

The conventional view of *Roe v. Wade* today is that the Justices embraced the pro-choice answer to the abortion issue. Indeed, "*Roe*" has come to stand for this proposition—but only in retrospect. *Roe* itself did not endorse the individual woman's right to choose for herself in the way that its holding is popularly understood today. Chief Justice Warren Burger, in his concurring opinion in the companion case to *Roe*, stated, "Plainly, the Court today rejects any claim that the Constitution requires abortion on demand."[3] Set against the dominant view of *Roe* today, this is an inexplicable observation. It may be, of course, that Burger misunderstood the import of the Court's ruling in which he joined; any careful student of Burger's judicial writings generally cannot dismiss this possibility. But in this case, Burger had it right.

If this is true, then there is a vast disjunction between what *Roe* has become and what it was at the moment in 1973 when it was decided. This may seem incredible; but, in fact, unless this disjunction existed, the Court's action in *Roe* would be even more incredible. In order to understand this proposition, we must do more than just look at the specific language of the Justices' opinions in *Roe*; we must try to reconstruct the jurisprudential and societal worlds of 1973 as the Justices saw them. In retrospect we can see that *Roe v. Wade* was, along with *Brown v. Board of Education*,[4] the most ambitious expression of judicial activism in this century. In terms of its explosive political impact, only the Supreme Court's mid-nineteenth century decision in the *Dred Scott*[5] case clearly exceeded *Roe*; *Dred Scott* precipitated the Civil War, but in some cases, this was only a more extensive and destructive civil war than the one *Roe* engendered.[6]

---

[3] Doe v. Bolton, 410 U. S. 179, 208 (1973) (concluding opinion).

[4] 347 U. S. 483 (1954).

[5] Dred Scott v. Sandford, 60 U. S. 393 (1857).

[6] See James Risen & Judy L. Thomas, <u>Wrath of Angels: The American Abortion War</u> (New York: Basic Books, 1998). This book's jacket blurb is, for once, not hyperbolic in its description: "Abortion has been at the emotional center of America's culture wars for a generation. Ever since the Supreme Court's landmark *Roe v. Wade* decision, abortion has in many ways defined American politics.... Above all, the twenty-five-year war over abortion has been

It may be that the Supreme Court that Earl Warren presided over would have been willing to engage in this kind of explosive law-making; such revolutionary zeal was the common complaint about the Warren Court among political conservatives. But by 1973, the Warren Court era had definitively ended. Warren himself had announced his resignation in 1968, just before Richard Nixon was elected President; Nixon had campaigned on an explicit platform promise to end the "judicial activism" represented by the Warren Court, and his first nominee to the Court as Chief Justice seemed, in light of his prior judicial service, to be perfectly fit to fulfill Nixon's promise. In fact, the new Chief Justice's full name was Warren Earl Burger, and the joke around Washington in 1969 was that Nixon could not have named a more suitable person to the Supreme Court to reverse everything that Earl Warren stood for.

Moreover, by the time the Court decided *Roe* in 1973, Nixon had appointed three more Justices. And yet, of these four Nixon appointees apparently pre-screened for their commitment to end the regime of "judicial activism," three concurred in *Roe*. Harry Blackmun wrote the Court's majority opinion, in which Lewis Powell as well as Burger concurred; only William Rehnquist dissented along with Byron White, a Kennedy appointee. Of the remaining Justices on the Court, three were certified activists from the "bad old" Warren Court days: William Douglas, William Brennan and Thurgood Marshall. This surviving remnant obviously did not constitute a working majority of the Court. Justice Potter Stewart, an Eisenhower appointee viewed generally as a "moderate" rather than activist judge also concurred in *Roe*. The stunning fact, overall, is that the Nixon "conservatives," these recently appointed anti-activists, were responsible for one of the three most immodest exercises of judicial authority in the history of the Republic.

When the four Nixon appointees were confirmed to the Court, no one thought to ask their views on abortion—and for a perfectly understandable reason. The judicial activism opposed by Nixon had been the Court's protective favoritism toward Blacks, accused and convicted criminals, and assorted dissenters such as Vietnam War protesters and avowed or suspected Communists. The Warren Court never adjudicated any case involving state abortion restrictions, and this absence is itself another surprising and puzzling aspect of *Roe*. The high point of Warren Court activism was undoubtedly *Brown v. Board of Education*, but the Court itself, for at least fifteen years before Earl Warren's appointment, had been repeatedly struggling with the status of Southern race segregation laws, obviously uncomfortable with their apparent

---

responsible for the most significant social phenomenon of our times—the political and cultural mobilization of Evangelical America. Furthermore, it has served as the lightning rod for the most intense and prolonged debate on the issue of separation of church and state since the founding of the nation."

constitutional legitimacy and yet hesitant about exercising judicial authority to strike them all down.[7] *Dred Scott* had also been preceded by significant Supreme Court decisions that anticipated its strong pro-slavery holding.[8]

In striking contrast, the first case ever to reach the Supreme Court regarding restrictive abortion laws came in 1971, just before its decision in *Roe*; and in that case, *United States v. Vuitch*, the Court was almost cavalier in its dismissal of the constitutional challenge. *Vuitch* involved a criminal conviction of a physician for performing an abortion in the District of Columbia; the physician claimed that, in permitting abortions only for "the preservation of the mother's life or health" and criminally punishing all others, the D.C. statute was unconstitutionally "void for vagueness." Justice Hugo Black, writing for a five-man Court majority, slid over the many ambiguities in the statutory formulation with this glib observation:

> Webster's Dictionary…properly defines health as the "state of being…sound in body [or] mind. Viewed in this light, the term "health" presents no problem of vagueness. Indeed, whether a particular operation is necessary for a patient's physical or mental health is a judgment that physicians are obviously called upon to make routinely whenever surgery is considered. We therefore hold that properly construed the District of Columbia abortion law is not unconstitutionally vague….[9]

Only two Justices, William Douglas and Potter Stewart, dissented from this conclusion, and Douglas alone invoked a woman's "privacy right" to control her own reproductive choices as the basis for his dissent. Stewart, as will be explored later, took a fundamentally different tack. Two Justices, William Brennan and Thurgood Marshall, refused to express any views on the merits of the statute on the grounds that the Supreme Court lacked jurisdiction to review criminal appeals from the District of Columbia court system. Justice Harry Blackmun agreed with Brennan

---

[7] See Robert A. Burt, *"Brown's* Reflection," 103 Yale Law Journal 1483, 1485-88 (1994).
[8] See Robert A. Burt, The Constitution in Conflict (Cambridge: Cambridge University Press, 1992) 172-86.
[9] 402 U. S. 62, 72 (1971).

and Marshall that the Court lacked jurisdiction, but stated that he would nonetheless express an opinion on the merits in order to resolve the deadlock produced within the Court. The four Justices had voted to uphold the abortion statute on the merits, two voted to overturn it and three (counting Blackmun) voted that there was no jurisdiction to address the merits. Blackmun then stated, without any elaboration, that he joined Justice Black's opinion on the merits, thus constituting a majority to reject the first constitutional challenge to abortion restrictions to have reached the Supreme Court. Here is yet another puzzle. Blackmun's position in *Vuitch* is, of course, an extraordinary oddity in light of his central role as the author of *Roe v. Wade* just two years after, striking down all restrictive abortion statutes.

All of these oddities—the absence of any Supreme Court attention to abortion restrictions until 1971, the absence of even a hint immediately before *Roe* from all but two Justices that the Court was concerned about the constitutional legitimacy of these restrictions, the recently ratified anti-activist credentials of the Court majority—combine to make *Roe* a seemingly inexplicable decision. The conventional account of *Roe* as a triumphalist statement of liberal individualism, of the right to privacy and a woman's right to choose for herself, makes sense only if we imagine the Court majority—and especially the three Nixon appointees who constituted the majority—as suddenly struck from the sky by a Constitutional Truth, rather like St. Paul falling from his horse. Sudden conversions can happen, of course. But that is not what happened in *Roe v. Wade*.

The first step in understanding what truly happened is to see that the Court did not embrace a woman's right to choose abortion. Chief Justice Burger's description, as mentioned before, had it right. Burger made his observation as a refutation of the dissenting claim by Justices White and Rehnquist that the Court's ruling meant that a pregnant woman could obtain an abortion for any reason she might choose—"convenience, family planning, economics, dislike of children, the embarrassment of illegitimacy, etc."—without any claim of a "threat to life or health." Burger countered that "the dissenting views discount the reality that the vast majority of physicians observe the standards of their profession, and act only on the basis of carefully deliberated medical judgments relating to life and health. Plainly, the Court today rejects any claim that the Constitution requires abortion on demand." Burger thus assumed that the medical profession would itself impose on women the same kinds of restrictive criteria on the availability of abortion as the state laws invalidated in *Roe*.

This appears to be a patently inadequate rejoinder to the dissenters' claim; even if the "vast majority of physicians" accepted these restrictions, state laws might still be needed to regulate the practice of the "small minority" of "excessively liberal" physicians—unless, of course, the Constitution prohibited any such state regulation. But this proviso simply restates the question: if the Constitution forbids such state regulation because a woman has a right to choose her own reasons for an abortion—whether she wants to protect her "life and health" or because, as the dissenters observed, she can't afford or just "dislikes" children—then Burger's rejoinder seems to miss the point. But in fact Burger's rejoinder identified, though only implicitly, the basic underlying premise of the Court's holding in *Roe*—that the restrictive state abortion laws violated the Constitution not because they interfered with a woman's privacy right *but because they infringed on physicians' rights to practice medicine, including the provision of abortions, as they saw fit.* This was the relevance of Burger's response to the dissenters, that the "vast majority of physicians...act only on the basis of carefully deliberated medical judgments relating to life and health." Even if there were a miscreant minority, the logic goes, this would be no justification for state regulation.

Justice Douglas also wrote a concurring opinion in *Roe*, and he explicitly based his concurrence on the existence of a woman's "privacy right" derived from the Court's prior decision in *Griswold v. Connecticut* overturning a state prohibition on the use of contraceptives by married couples. This was also the basis for Douglas's dissent in *Vuitch* two years earlier. But in *Roe*, as in *Vuitch*, Douglas was alone in giving this primacy to a woman's right to choose abortion. Justice Stewart had been the other dissenter in *Vuitch*, but he had a very different basis for his dissent. Stewart asserted that the criminal conviction under the District of Columbia statute was invalid on the ground that

> the question of whether the performance of an abortion is "necessary for the ...mother's life or health" is entrusted under the statute exclusively to those licensed to practice medicine, without the overhanging risk of incurring criminal liability at the hands of a second-guessing lay jury. I would hold, therefore, that a "competent licensed practitioner of medicine" is wholly immune from being charged with the commission of a criminal offense under this law.[10]

There is some ambiguity in Stewart's position, whether he found physician immunity based on his reading of the D.C. statute or on his construction of a constitutional right to physician immunity. But there is no such ambiguity in Justice Blackmun's opinion for the Court in *Roe*. At the end of his lengthy and somewhat rambling exposition, Blackmun states this conclusion: "For the period of pregnancy prior to this 'compelling' point [the end of the first trimester], the attending

---

[10] United States v. Vuitch, 402 U. S. 62, 96-97 (1971).

physician, in consultation with his [sic] patient, is free to determine, without regulation by the State, that, in his medical judgment, the patient's pregnancy should be terminated." Though earlier in his opinion Blackmun had referred in general terms to a "right of privacy" as "broad enough to encompass a woman's decision whether or not to terminate her pregnancy," he clearly emphasizes by the end of his opinion that this was not the central justification for invalidating the state abortion restrictions. It was not the woman's decision, but her physician's, which was at stake in no uncertain terms.

Thus, immediately after his endorsement of the (presumptively male) attending physician's freedom to choose, Blackmun reiterated:

> To summarize and to repeat: A state criminal statute of the current Texas type, that excerpts from criminality only a life-saving procedure on behalf of the mother, without regard to pregnancy stage and without recognition of the other interests involved, is violative of the Due Process Clause of the Fourteenth Amendment: (a) For the stage prior to approximately the end of the first trimester, the abortion decision and its effectuation must be left to the medical judgment of the pregnant woman's attending physician.[11]

After listing the permissable range of state regulation in the second and third trimesters, Blackmun again returned to his central premise:

> [Our] decision vindicates the right of the physician to administer medical treatment according to his professional judgment up to the points where important state interests provide compelling justification for intervention. Up to those points, the abortion decision in all its aspects is inherently, and primarily, a medical decision, and basic responsibility for it must rest with the physician. If an individual physician abuses the privilege of exercising proper medical judgment, the usual remedies, judicial and intra-professional, are available.[12]

With this stirring endorsement of "the right of the physician," Blackmun concluded his opinion; he immediately proceeded to the formal declaration that "the Texas abortion statutes, as a unit, must fall" and remanded the case to the lower court for enforcement proceedings. The pro-choice celebrators of *Roe v. Wade* focus exclusively on Blackmun's earlier references to the "woman's decision whether or not to terminate

---

[11] 410 U. S. at 165.
[12] Id. at 165-66.

her pregnancy;" they ignore the clear structure and repeated emphasis in his opinion that the physician's right, not the woman's, was at the center of the Court's holding.

In the companion case to *Roe*, involving the more liberal Georgia statute which permitted abortions to protect the mother's physical and mental health, as well as her life, the Court was equally explicit in placing physicians' rights at the center of its reasoning. The Georgia statute required individual physicians to obtain approval of any abortion by two additional practitioners and then by a three-member hospital committee. The Court then held that this procedural requirement had

> no rational connection with a patient's needs and unduly infringes on the physician's right to practice. The attending physician will know when a consultation is advisable—the doubtful situation, the need for assurance when the medical decision is a delicate one, and the like. Physicians have followed this routine historically and know its usefulness and benefit for all concerned. It is still true today that "reliance must be placed upon the assurance given by [the physician's medical] license, issued by an authority competent to judge in that respect, that he possesses the requisite qualifications." *Dent v. West Virginia*, 129 U. S. 114, 122-123 (1889).[13]

This citation of an 1889 Supreme Court precedent is quite revealing of the *Roe* Court's essential mindset. 1889 was, of course, an era not only of the high point for Victorian paternalism but specifically a moment when a visible alliance was forged between medical authority and state power. In *Dent v. West Virginia* itself, the Court unanimously upheld a criminal conviction under a recently enacted state medical licensure law against a practitioner "publicly professing to be a physician" but without having graduated from a "reputable medical college." In this struggle between folk practice and the organized proponents of scientific medicine, the Court clearly sided with Science: "[Medicine] has to deal with all those subtle and mysterious influences upon which health and life depend.... Every one may have occasion to consult him, but comparatively few can judge of the qualifications of learning and skill which he possesses."[14]

This historical moment when state licensure of medical practice took hold was the same time when state laws restricting abortions were first enacted; until the late nineteenth century, abortions were freely practiced by midwives and other folk practitioners just as purveyors of patent medicine and other self-styled healers freely competed for public

---

[13] Doe v. Bolton, 410 U.S. 179, 199 (1973).

[14] 129 U. S. at 122.

allegiance. As Paul Starr observes in his excellent study, <u>The Social Transformation of American Medicine</u>, it was only in the late nineteenth century that dependence on professional authority became the dominant cultural response to public concerns about health and life. The relatively low status (and incomes) of American physicians before the Civil War "stemmed partly from Americans' ingrained self-reliance, their disbelief in the value of professional medicine, and the ease with which competitors entered the field."[15] By the end of this nineteenth century, however, that ethos of democratic individualism had been fundamentally displaced: "Americans became willing to acknowledge and institutionalize their dependence on the professions.... On the shoulders of broad historical forces, private judgment retreated along a wide frontier of human choice."[16]

In *Roe v. Wade*, the Supreme Court struck down state licensing restrictions regarding abortion—but not in order to vindicate "private judgment," not to unravel Americans' "dependence on the professions." The explicitly avowed goal of the Court majority was to strike down state laws in order to enhance professional authority—to immunize medical practitioners, as Justice Stewart had put it in *Vuitch*, from any "risk of incurring criminal liability at the hands of a second-guessing lay jury." As Justice Blackmun stated at the end of his opinion in *Roe*, "[Our] decision vindicates the right of the physician to administer medical treatment according to his professional judgment." The underlying impetus for *Roe* was deeply conservative—to reassert the traditional dominance of professional judgment in the face of challenges to their caretaking authority.

This essential conservatism explains what is otherwise so puzzling about *Roe*: how it came about that Richard Nixon's appointees formed the core of the *Roe* majority. It also explains the extraordinary lurching of the Court from its casual dismissal in 1971 of the constitutional challenge in *Vuitch* to its wholesale invalidation of abortion restrictions just two years later. This was a moment of extraordinary tumult in American social life; every embodiment of traditional order and authority was under siege. Popular disaffection toward the Vietnam War was at its height; not since the Mexican-American War in 1848 or perhaps the darkest days of Northern military setbacks in the Civil War had the American public so widely disclaimed support for its established military leadership in battle. Domestic disorder also appeared frighteningly dominant in the decade before 1973: the assassination of President Kennedy, Martin Luther King, Jr. and Robert Kennedy, and the crippling attack on George Wallace; the spread of race

---

[15] Paul Starr, <u>The Social Transformation of American Medicine</u> (New York: Basic Books, 1982) 65.
[16] Id., 17, 140.

riots and the rise of Black militancy; the explosive civil rights and anti-war protests on college campuses across the country.

This was also a time of unusual instability in the institutional life of the Supreme Court. In the short course of three years, four new Justices took office following Earl Warren's retirement, Abe Fortas's forced resignation, and the deaths of Hugo Black and John Marshall Harlan. The Court itself was institutionally off-balance and visibly struggling to respond to the extraordinary social turmoil outside. Abortion was not the only subject that unpredictably erupted in the Court at this time. The Court's treatment of the death penalty was equally sudden, convulsive and erratic: from the almost casual dismissal of any constitutional challenge in one of the last opinions by Justice Harlan in 1971 to the abrupt, utterly unexpected invalidation of all death penalty statutes just one year later, then followed in 1976 by an equally abrupt retreat and reaffirmation of the legitimacy of capital punishment.[17] In both the death penalty and abortion cases, it appeared as if the Court was drawn into suddenly eruptive disputes about the legitimacy of traditional social orderings and that almost in spite of itself the Court could not simply ratify existing arrangements but felt forced to alter them in some way so as to bolster their perceived legitimacy. Especially regarding abortion, the Court's majority effort to immunize physicians' authority from any lay control—whether by criminal law juries or by elected legislatures—was not a radical embrace of liberal individualism. *Roe v. Wade* was originally conceived as a counter-revolutionary tactic by embattled, off-balance social conservatives.

Viewed in this light, *Roe* expressed the same strategy that organized medicine had followed in pressing for reform of state abortion laws in the 1960s. As Kristin Luker has described in her superb sociological study of the politics of abortion in California, the state law restrictions as adopted in 1872 had customarily been understood to exempt "reputable" physicians in their freedom to exercise independent judgment, but this implicit understanding was breached by a few high-visibility prosecutions. "In short order, between 1964 and 1966, the American Medical Association, the American Bar Association, the American Academy of Pediatrics, the California Medical Association, the California Bar Association, and numerous other [professional] groups threw their support behind abortion reform."[18] In 1967, the California legislature enacted a new abortion law which kept intact the regulatory structure for hospital committee approval but liberalized the standard to protect against "grave impair[ments]" of the pregnant woman's physical or mental health. Ironically enough, in light of his subsequent fervent

---

[17] See Burt, supra note 8, 331-38.
[18] Kristin Luker, <u>Abortion and the Politics of Motherhood</u> (Berkeley: Univ. of California Press, 1984) 88.

identification with the pro-life forces following *Roe v. Wade*, this liberalized abortion statute was signed into law by the then-Governor of California, Ronald Reagan. Even more ironically, the practical effect of this new law was to establish a de facto rule of abortion on demand. Luker's account is worth quoting at some length on this score:

> In 1968, the first full year under the new law, 5,018 [recorded] abortions were performed. In the next year, however, the number tripled, to 15,952. The following year that number itself quadrupled, and 65,369 abortions were performed. In 1971 it almost doubled again, and 116,749 abortions were performed. In 1972 the rate stabilized at a little over 100,000 abortions and has remained at that level to the present. In four short years, therefore, the number of abortions sought and performed in California increased by *2000 percent*. Moreover, by 1970 it was becoming apparent that what had been proposed as a "middle-way" solution had in fact become "abortion on demand." It is possible that the mechanisms of medical review (and psychiatric review in the case of those using the "mental health" criterion) may have been sufficiently cumbersome and expensive to discourage some women for applying for an abortion in the first place; but by late 1970, of all women who applied for an abortion, 99.2 percent were granted one. By 1971 abortion was as frequent as it would ever become in California [even after the Court's decision in *Roe v. Wade*] and one out of every three pregnancies was ended by a legal abortion. ...By 1971, women in California had abortions because they wanted them, not because physicians agreed that they could have them.[19]

It is crucial to understand that this extraordinarily liberalized regime was, on its face, still subject and still deferential to the traditional caretaking authority of physicians. The momentary eruption of visible social conflict about the relationship between lay judgment and the proper range of physicians' authority was effectively reburied. Although in actual practice, physicians deferred more readily—indeed, it appears virtually always—to pregnant women's wishes, nonetheless women still engaged in public rituals of deference toward physicians' judgments. Traditional caretaking arrangements still intertwined with traditional notions of deference to socially constituted authority were reaffirmed at the same time that challenges to these arrangements were quietly appeased.

---

[19] Id., 94.

The conservative Supreme Court majority in *Roe v. Wade* instinctively reached for the same resolution. But the center would not hold, as the poet William Butler Yeats observed of an earlier socially disrupted era when "mere anarchy is loosed upon the world."[20] Though the Court itself has altered its vocabulary in its repeated engagements with the abortion dispute since *Roe v. Wade*, and its initial emphasis on physicians' rights has vanished from its justifications striking down state restrictions, the current Court majority continues to interweave claims for deference to traditional caretaking authority with its endorsement of women's individualist claims to control the abortion decision. In 1992, four Justices were prepared to overrule *Roe v. Wade*. The plurality, however, who formed the core of the narrow Court majority reaffirming *Roe* were virtually explicit in subordinating the individual woman's right to the Justices' own view of the social necessity for such deference. In this instance, however, these Justices were not commanding to physicians' authority; they were honoring the Court's own institutional authority. In their joint opinion, Justices O'Connor, Kennedy and Souter stated,

> In 1973, [the Court] confronted the already-divisive issue of governmental power to limit personal choice to undergo abortion, for which it provided a new resolution based on the due process guaranteed by the Fourteenth Amendment. ... [I]ts divisiveness is no less today than in 1973, and pressure to overrule the decision, like pressure to retain it, has grown more intense. A decision to overrule *Roe's* essential holding under the existing circumstances would address error, if error there was, at the cost of both profound and unnecessary damage to the Court's legitimacy, and to the Nation's commitment to the rule of law. It is therefore imperative to adhere to the essence of *Roe's* original decision, and we do so today.[21]

As with *Roe* itself, this is a deeply conservative response to perceived social turmoil. In 1992 the Court plurality explicitly acknowledged what the conservative majority in *Roe* only dimly perceived: that in trying to reassert traditional authority, the Court was mounting the back of some "rough beast" (to quote Yeats again) and, as the Justices plainly saw in 1992, the risk was high that the beast would turn to swallow its riders. The true identity of this beast is not, however, the firestorm of protests

---

[20] "The Second Coming," in <u>Collected Poems of W.B. Yeats</u> (New York: Macmillan, 1958) 184.
[21] Planned Parenthood of Southeastern Pennsylvania v. Casey, 505 U. S. 833, 868-69 (1992).

for and against freely available abortion that has convulsed our national politics since 1973; neither is it the proclaimed radical individualism of the pro-choice forces nor the proclaimed radical communitarianism of their pro-life opponents. These combatants are themselves trying to master the unruly, frightening anarchy implicit in a world where neediness finds no reliable source of social support, where the weak expect to be abused rather than nurtured by the strong.

Whatever the possibility that this deep source of social discontent might be salved, it seems unlikely that much sustenance will come from any of the answers proffered in our current national polemics—not from the Supreme Court's demands for renewed deference to traditional caretaking authority, nor the pro-choice pretense that each individual can find nurturance enough from her own independent choices, nor the pro-life insistence that in attending to the vulnerability of some, we should blind ourselves to the equal vulnerability of others. The real question may not be how this urgent vulnerability can be successfully met. It may instead be how we can learn to live with unsatisfied vulnerability, to accept inherent conflict between one person's needs and another's, to endure in the face of tragedy. In the current abortion debate, all sides are making universalistic claims that tragedy can be avoided, that all neediness can be successfully satisfied. The Supreme Court's repeated insistence that the abortion dispute can be resolved by an appeal to constitutional principle is one such false universalistic claim.

HOME SWEET HOSPITAL:
THE NATURE AND LIMITS OF FAMILY
RESPONSIBILITIES FOR HOME HEALTH CARE
Carol Levine

"As the eldest of five living children, I took responsibility because the others wouldn't. It was a decision which cost me my health, my job and more than likely, my marriage."
—Family caregiver, Georgia (1993)

"She is my mother; she is my responsibility; she is my blessing."
—Family caregiver, Georgia (1993)

"Families need the opportunity to take responsibility for themselves."
—Vice-President Albert Gore, at conference "Family Reunion II: Reinventing Family Policy," Nashville, TN (1994)

"A primary assumption that runs through this history [of long-term care] is that families, a euphemism for wives and daughters, would take primary responsibility for their disabled or impoverished elders."
—Martha Holstein and Thomas Cole (1995)

"Most home health care agencies require there to be a 'responsible person' who can step in and personally provide or arrange for coverage if the aide is ill, detained...or irresponsible....The responsibility of the involved family member is nearly pervasive."
—Nancy N. Dubler (1990)

"Responsibility" for the care of one's family members is, according to these selected quotations, an opportunity that families need (and presumably lack), a task traditionally assigned to women, a personally costly obligation, a loving act, a total commitment. Responsibility is, variously, something to cherish, something to resent, something that is thrust upon one, or something that one tries to thrust upon others.

This essay is about concepts of family, private and public responsibility, and family capacity to provide care. Although informed by existing data and the work of pioneering researchers, it is not limited to what is quantitatively known or perhaps even knowable. And although it draws upon principles of biomedical ethics, it applies them in the arena of the family rather than the individual. The essay is intended to look broadly toward future home care policies and practices through the special lens of family or "informal" caregiving.

As the primary caregiver for my husband, who was totally disabled as a result of traumatic brain injury in an automobile accident nine years ago, my experiences also inform this essay. Personal experience is both a powerful reality check and a potential source of bias, but I use it as only one source of information. No individual's or family's story stands for the vast range of experience, both positive and negative, of caregiving in the United States today. Caregiving is experienced differently by, for example, the partner of a young man with AIDS; the husband of a 35-year-old mother of three with multiple sclerosis; the parent of an independent and resourceful 25-year-old with a spinal cord injury; the daughter of a frail elderly woman who lives alone but needs assistance in shopping, transportation, and financial management; the son of a woman in a hospice program who is dying of brain cancer; and the wife of a man with Alzheimer's disease whose behavior is erratic and often hostile.

Although typically considered part of the private realm of intimate relationships, family caregiving is greatly influenced by the cultural, political and economic context of American society. Family caregiving is a traditional response to illness and disability, but it is not a static phenomenon. Some current problems have historical roots. For example, the burden of family care has always fallen especially on women, and most especially on poor women. The tension between paid work and family care felt so keenly today has existed in the United States at least since the nineteenth century.[1]

Currently both families and health care are changing dramatically. While in earlier eras, some individuals lived to great old ages, the average life expectancy was decades less than it is now. In 1850 just over 2 percent of the population was over the age of 65; now the percentage of people over 80 is growing rapidly.[2] Although being old does not necessarily mean being frail or ill, there is an increase in diseases of aging, especially Alzheimer's disease. Because physicians had few effective treatments until the antibiotic age, most people who suffered severe trauma or serious illness either got better or died. Nature put a limit on caregiving. In the twentieth century the advent of scientific medicine and the benefits of research, public health measures, better nutrition and safer jobs have enhanced and extended lives. Moreover, some recent successes of acute care medicine—for example, in the care of newborns and trauma patients—have also created a population of adults dependent to

---

[1] Abel 1995.

[2] Institute for Health and Aging at the University of California, San Francisco, 1996.

unprecedented degrees on technology and on other people for basic survival.

Home care of chronically ill or disabled people in the twenty-first century will take place under vastly different conditions than in the nineteenth century or even in the mid-twentieth century. Old assumptions and patterns of care do not fit present-day realities. Home care will be affected by the needs of an aging population, changes in family and household organization, epidemiology, women's increasing participation in the labor market, and, perhaps most decisively, the ongoing trend toward a health care delivery and financing system that uses hospitals and professional and public resources sparingly and patients' homes and family caregivers liberally.

The process of market-driven health care is still evolving, but trends are clear. Health care costs are being constrained through reduced length of hospital stays, increased outpatient and community-based care, and reductions in home care benefits available through insurance, managed care organizations, or public programs. Individuals and families will be under increased pressures to pay more direct costs; families will be expected to provide more hands-on, often technologically complex care; undertake greater burdens for longer times; and forgo more educational, career, and social opportunities. The SUPPORT study of decision-making at the end of life has documented the devastating financial impact on families of terminally ill patients, even when they had private insurance (Covinsky et al., 1994). Moreover, most families do not want and cannot afford to place their relatives in long-term care facilities; only five percent of the elderly are in nursing homes. Nevertheless, the human and social costs of maintaining patients at home are very high. Some families respond heroically to daunting challenges of care; without diminishing their sacrifices, it is unrealistic and unfair to hold all families to a standard of heroism. The ultimate result of doing so will be higher rather than lower social, health care, and other costs.

The paradigm of the past decades—the elderly, retired (or never employed) woman readily providing assistance to an even older and frailer family member—is only one significant part of the home care picture. Depending on the definition of caregiving and the population surveyed, estimates of the number of American caregivers range from 7 million to 27 million.[3] The majority are indeed middle-aged or elderly women. Nevertheless, changing patterns of illness (AIDS, for example) and family structure are bringing more men and younger people into these roles.[4] As the population becomes more culturally diverse, there will also be more diverse norms of caregiving.

---

[3] National Alliance for Caregiving, 1997; Manton, Corder, and Stallard, 1993.
[4] Marks 1996.

We approach the twenty-first century without a coherent, consistent, and comprehensive long-term care policy or a coherent, consistent, and comprehensive family policy. Long-term family caregiving policy is a vacuum within a vacuum. A basic question is: How should public and private responsibilities be allocated at the family, individual, and community level? This leads to several further questions:

1. What are the justifications for looking to families to take responsibility for ill, elderly, and disabled relatives?
2. If family caregivers take on significant responsibilities for a relative's care, whether willingly or reluctantly, what are the limits of their responsibilities?
3. What are the relationships between public and private responsibilities for home care?
4. How can the capacity of specific families to undertake significant levels of home care be assessed and addressed?

I. Concepts of family responsibility: Justifications and limits

Two major concepts—"family" and "responsibility"—are commonly linked but seldom defined, either singly or jointly. Much of the rhetoric that places primary, often sole, responsibility for caregiving on families assumes an unvarying standard: Families have always done it, so they should always do it, and it is nobody's problem but theirs. Underlying these assumptions is another, even more basic one: Only people who are closely related by blood, marriage or adoption count as family.

While family is a basic organizing structure of all human societies, definitions of family have varied throughout history and by culture. Even in cultures where family is the primary unit of attachment, communities may supplement what families can provide or assume some caregiving roles, such as helping new mothers or caring for the dying. Religious institutions have a long history in providing care for the ill and dying; modern hospitals grew out of this tradition. Until the 1970s and 1980s in the United States people with psychiatric problems, mental retardation, severe disabilities, and other conditions were routinely institutionalized. The shortcomings of care, abuses of human rights, and economic costs of the past are powerful deterrents to a return to institutions for these populations. Still, even in our own time, families have not been expected to provide all necessary care.

American views about "family" are often conflicting and contradictory. Powerful political and legislative efforts supported by public opinion are attempting to circumscribe collective or social responsibility for supporting families with children, especially female-headed households. At the same time families are generally believed to be extremely fragile and disintegrating institutions. Families are expected to do more and more but are believed to be capable of less and less.

For the purposes of this essay, my working definition of family is broad but not unlimited. "Family members are individuals who by birth, adoption, marriage, or declared commitment share deep, personal connections and are mutually entitled to receive and obligated to provide support of various kinds to the extent possible, especially in times of need."[5] Written from the perspective of the impact of AIDS on families, this definition is equally applicable to other circumstances of family caregiving. It respects traditional notions of family and recognizes nontraditional forms of commitment. It acknowledges the power of biological ties but also stresses the bonds of voluntarily chosen relationships.

This working definition will undoubtedly be problematic at the boundaries of who is included and who excluded; the central core of deep, long-term, emotional commitment should hold firm. From the home care perspective, this broad definition implies both that families may have more people as resources than might be assumed by asking only about the nearest biological kin and that preconceptions about who "should" provide care may be counterproductive.

Implicit in this definition is a sense that family members bear some moral, if not necessarily legal, responsibility for each other's welfare, as well as a recognition that there are limits to that responsibility. What is the source of this responsibility? Although there is a voluminous literature on family caregivers, there is only a limited literature on why responsibility for care falls first or only to families, how families should balance competing responsibilities, or when responsibility moves beyond "virtue" into the superogatory realm of martyrdom.

At its most literal meaning, "responsibility" is a response or a reply. In his book *A Grammar of Responsibility*, Moran points out that "responsibility begins with an aural/oral metaphor: first one listens, then one answers...." According to this view, the responsible person is "someone who listens, recognizes the word as a personal address, and is impelled to answer." Until the nineteenth century, the primary sense of responsibility was "to" something—God and divine judgment, in particular. In the late nineteenth century, with the advent of liberal political thought, the sense of being responsible "to" waned, and being responsible "for" things became explicit. This sense of responsibility has dominated modern thinking, but it has, Moran believes, been misunderstood. One is not responsible "for" one's life, as is frequently stated, but for one's acts or failures to act. "Rights" are often presented

---

[5] Levine 1991.

as requiring the balance of "responsibilities;" rights should, Moran claims, be balanced by duties and obligations, which grow out of the process of responsibility.[6]

Jonsen, for example, defines responsible persons in terms of fulfilling their obligations: "[They] know what needs to be done, they appreciate its significance, they proceed on their own, they get the job done, and they do it well."[7] While this aptly describes what one would like to see in a public official, a well-trained professional, or a colleague at work, when applied to family caregivers it is only a partial description and only applicable to those who have struggled, largely on their own, to develop the skills and mindset to fill the role. Family caregivers, even with the best of intentions, are not "responsible" in this sense of obligation. They are not responsible "for" their ill relative but for actions fulfilling their duty to that person, or if the person is unable to act for himself, for actions taken in his or her stead.

Other, more legalistic senses of responsibility relate to causation or liability, such as being responsible for causing an accident or liable for the resulting damages. Responsibilities can fall both to individuals and to institutions. Many of the most common senses of responsibility have to do with formulations of contract, consent and accountability. Rules and regulations, while subject to interpretation, spell out the limits of legal responsibility.

Family responsibility is a more amorphous concept, with fluid boundaries and interpretations. It may derive from religious teachings, cultural tradition, emotional bonds, gratitude for past acts, or a sense of obligation apart from love. In the health care setting family responsibility is triggered by a family member's need for care. Some of the ways in which the family may respond are emotional support, surrogate decision making, economic contributions, care management, and hands-on care. Perhaps the most important justification for looking first to families for caregiving is that most families themselves want, or at least accept, this responsibility. Taking care of each other comes with being a family.

The definition of who counts as family has another dimension for home care. When "formal" or paid caregivers such as home health care aides are involved extensively in care, their relationship to the patient and family may test the boundaries of definitions of "family" and expectations of what family should provide. There are many different living situations and types of interactions between aides and family members, some infrequent and casual, others intense and prolonged. There is a tendency for workers' advocates to stress the potential for families to exploit workers, and for family members to emphasize their

---

[6] Moran 1997: 36-37.
[7] Jonsen 1968.

experience with unreliable, incompetent or abusive workers. Both types of problems exist. Patients, families, and aides may have difficulty in adjusting to each other, and very seldom obtain the necessary training or limit-setting that would be helpful in preventing misunderstandings.

Family members may be responsible for hiring and supervising the aide, a managerial role outside typical family relationships. They may ask aides to provide services beyond their assigned duties, as they might do with family members, creating resentment and resistance on the part of the aide. On the other hand, aides may view themselves as "part of the family," and may in fact be, better caregivers or more devoted to the patient or respectful of his or her autonomy than family members. Aides may subtly or not so subtly undermine already strained family relationships which will persist past the aide's tenure.

Although each situation varies, it is probably best for family caregivers and patients to develop a relationship with aides that is informal and collaborative, even affectionate, but not to cross boundaries that invade the patient's and family's privacy. Privacy does not just include space but also marital relationships, financial affairs, or other matters that do not directly affect the client's health and well-being. The worker should, of course, report serious examples of family abuse to supervisors or other authorities.

A common label applied to modern families is "dysfunctional;" this may mean that some members of the family, or all of them, have antagonistic relations among themselves and with outsiders, that they engage in illegal behavior, or just that they are atypical in structure, attitudes, or actions. In the medical context the label of dysfunctional is often given to families who do not conform to institutional rules, who disagree among themselves, or who disagree with medical recommendations. Dealing with such families is difficult for medical professionals; a recent Harvard workshop on end-of-life care lists "dealing with dysfunctional families" as an "advanced topic." Some families are in fact dysfunctional in a psychological sense; others are led to antagonistic or counterproductive behavior by the crisis of illness. However, even functional families can be overwhelmed by the hospital environment and the attitudes and behaviors of staff, which, to the family, can seem truly dysfunctional.

Family caregiving involves intimate relationships, a private setting, no exchange of money (although reciprocal services and emotional debts are common), and an almost total lack of outside regulation. The vast majority of family caregivers assume their roles as a normal and natural, if often frustrating and burdensome, part of life. They may not even recognize that they are "family caregivers," seeing themselves primarily as wives or husbands, sisters or brothers, daughters or sons, lovers or companions. They meet the classic definition of "role responsibility" (having an obligation to do something because of who one is in relation to another person). For the most part they do not want to abdicate their private responsibility to the public sector, which is often viewed as impersonal, uncaring, and bureaucratic. They may view their privacy as more important than the potential benefits of formal home care. Many do, however, need and want support and assistance to the degree and of the kind that they themselves deem important and can control. The fact that families often take the first step toward accepting responsibility, however, does not mean that this decision removes all responsibility from others or that it is irrevocable and interminable.

Underlying various views of what families should be held responsible for are at least two value-laden views of families. One view holds that families are important ends in themselves and should be highly valued and supported through societal norms, various supportive measures, and economic incentives. Families give meaning and depth to human relationships. Membership in a family by its very nature also creates obligations. As philosopher Patricia Smith puts it, "Family obligation is potentially the most burdensome—depending on the nature of the particular family membership—and the most beneficial obligation human beings can have."[8] In the health care setting, this view of families leads to the conclusion that families should be given equality in decision making, or at least full consideration of their interests. The interests of the patient and the family—and especially those who provide the greatest levels of care—may conflict. For example, a daughter may not want to provide end-of-life care to the parent who abused her in childhood and who demands her participation in care now. On the other hand, in the same situation, the parent may not want the daughter's care but the daughter may insist on providing it as a means of reconciliation or as a way to obtain control at last. In these cases patient autonomy, an essential value in health care in terms of refusing or accepting treatment, is not absolute, but must be tempered by the impact of choices on family caregivers.

Another view is that the family in the health care setting has instrumental value; that is, it is important mainly for what it can provide to the patient. In this view the family's ability to provide and sustain home health care is important to the extent that it benefits the patient. It is also important from an economic view because it is a significant resource that cannot easily be replaced by paid, professional care. Physicians and other health care providers have largely adopted this view of families; it is supported by the basic professional obligation to serve the patient's interests, and it avoids dealing with complex relationships within families. There are, to be sure, exceptions (as within pediatrics and family practice) but professional norms and practice largely follow this view.

---

[8] Smith 1993.

While there is ample evidence that millions of American families accept extremely onerous levels of responsibility, and the vast majority do not abandon their relatives, some families do not provide care. In the United Hospital Fund's New York City Medicaid home care survey, 40 percent of the care recipients received assistance with personal care or household chores regularly from their families.[9] Nearly a third of these caregiving families provide more than 40 hours of care a week. On the other hand, one third of care recipients live alone and get no assistance from family members with personal care or household chores. Nearly half of those over 85 years of age live alone. What are the reasons some family members avoid responsibility for care? Is it the economic burden, the prior relationship with the patient, geographic distance, unwillingness to give up personal freedom or career goals? A better understanding could make it possible to develop interventions that might make the difference for some family members between being able to do some or all of this task and not doing it at all.

For some families the problem is accepting responsibility, for others equitably dividing responsibility. For many, however, the problem, particularly for the caregiver most involved, is a difficulty in recognizing limits to responsibility. Some caregivers come to believe that no one else can provide the care as expertly as they can; many, particularly elderly, patients refuse to accept care from "strangers." In addition to avoiding the myths of the idealized past and the idealized family, it is important to avoid idealizing the patient. Some patients turn their illnesses and disabilities into opportunities for expressing love, gratitude and spiritual growth. Others become demanding, hostile, and physically or emotionally abusive, even and especially to their caregivers. Jane Bendetson, who cared for her husband for 23 years, writes of her own and others' anguish: "I am I. I am more than an adjunct to a disease ... more than hands to lift, to bathe, to assist ... more than a body to feed him, to phone for help, to hold when the nights are hard ... more than an object against which to rage, to demand of, to be there whenever, wherever ....."[10]

While respite services are probably the most frequently suggested option for exhausted family caregivers, some caregivers are not willing to accept even this type of assistance. Some respite programs, such as the California state-sponsored program, are successful, and there is a long waiting list. Other creative efforts are needed to prepare caregivers to accept respite and to provide it in ways that both caregiver and patient find satisfactory.

Beyond day-to-day limits of caregiving, there are limits to what a family can provide at home, particularly over long periods of time.

---

[9] Hokenstad, Ramirez, Haslanger, and Finneran 1997.
[10] Bendetson 1997.

Writing about Alzheimer's disease, ethicist Stephen Post declares, ."..the caregiving within the family is a precious moral resource—so precious that it should not be exhausted.... Given the frequent absence of support for caregivers, we must be tolerant of those who are unable to handle the stress of stewardship and therefore must relinquish direct care.... No person is morally reprehensible for having failed to do something that is virtually impossible, no matter how strong his or her character."[11] Nelson and Nelson agree that burdens are unjust if "they importantly compromise the core functions of family; or they importantly compromise core projects of individuals; so long as good faith efforts to share, negotiate, and reduce burdens have been made."[12] Nevertheless, following one of their core principles—"family members are stuck with each other"—they believe that even unjust burdens of care may have to be provided, "because there is no one else to do it." This conclusion is unsatisfactory on two grounds: it diminishes the argument based on justice and it provides a convenient rationale for failing to assist overwhelmed caregivers.

In such cases nature will, as in the past, define some limits to caregiving. But it will be the caregiver, not the patient, who succumbs. Admittedly, defining limits of responsibility is difficult, and criteria will vary. An inability to continue caregiving, whatever the reason, should not be considered a failure, either by the rest of the family, medical or social service providers, or society. It is unjust and unrealistic to expect caregivers to forgo other obligations, to their children, for example, and to give up so much of life's satisfactions and challenges that they lose not only their health and financial security, and their relationships with others, but also their identity as persons. Total self-sacrifice may be ennobled in legend; it is a decidedly unsatisfactory way to live one's life and a poor basis for public policy.

Just as family members have moral obligations toward each other, society has moral obligations toward families. If families truly are the core institutions embodying critical human values, then society has to ensure that caregiving does not destroy them.

II. Concepts of responsibility: The public/private distinction

Social or public responsibility—always an ill-defined and ambiguous concept in American life—is undergoing radical redefinition in terms of public welfare programs, indigent health care, social services for legal immigrants, and more. The thin veneer of consensus that has supported some sense of communal responsibility in this century is cracking. For many, privatization is the paradigm, corporate control the mechanism, and diminished government involvement the desired result.

---

[11] Post 1995.
[12] Nelson & Nelson 1995.

Although the sectors variously labeled public, private, corporate, professional, and governmental have distinctive characteristics, they are also interrelated. Decisions made in each sector profoundly affect the others. Underlying most discussions of public policy is the assumption that it affects only those people who are eligible for the program—those who are elderly, poor, disabled according to Social Security Administration criteria, veterans of the armed forces, children of working poor parents, etc. In fact, public policies regarding home care have a powerful impact on the private sector.

Explicit public responsibility for home and community-based services has largely been confined to the elderly and the indigent, and with the singular goal of keeping patients out of nursing homes. Concern about family caregiving has wavered between policies intended to be supportive of families so that they maintain their relatives at home and those that cut costs by expecting more of families. According to Stone and Keigher:

> The confusing array of long-term care policies that presently exist actually provide little tangible support to family caregivers. They have largely functioned to deter the abrogation of family responsibilities by helping caregivers only marginally and only when certain consequences with negative cost implications for the government are imminent.[13]

The ambivalence about families is seen currently in several policies. The Family Medical Leave Act, for example, provides for up to 12 weeks of unpaid leave for family health needs for employees in firms with over 50 employees. This policy can be very useful for new parents who have made financial plans to cover the birth or adoption of a child, but it is not very helpful to the spouse caregiver who is the sole source of family support and cannot afford to take unpaid leave. A provision in the Health Insurance Portability and Accountability Act (1996), later revoked, criminalized transfer of assets to become eligible for Medicaid, under the banner of "rooting out fraud" and "cost containment." Hospice, as an option to aggressive end-of-life care, is covered by Medicare (as long as there is a primary caregiver, typically a family member); these services offer considerable family support. Yet, the federal Health Care Financing Agency (HCFA)'s Operation Restore Trust is seeking to recover funds from hospice programs that admitted patients "too early," that is, patients who live beyond the covered six months. Families must cope with these dying patients without hospice or other support.

---

[13] Stone & Keigher 1994.

What the public sector will pay through Medicare or Medicaid or other government programs is a major factor in determining how private enterprises develop and price services and products. While Medicaid and, in particular, Medicare are high on the policy cost-containment agenda, and considerable research has been conducted on the recipients, little public policy attention has been paid to the large numbers of people needing extensive care who are not covered by Medicare or Medicaid and who are either uninsured or whose private insurance has extremely limited home care benefits. Middle-income, employed people who need to supplement their own caregiving with formal care find that services through home care agencies are often unaffordable. Agencies have few incentives to deal with individual clients when most of their business comes from institutional payers, primarily the government or large managed care organizations. As a result an underground economy in home health care has grown up, with no quality control or supervision other than that provided by the usually untrained and overburdened family member.

As already noted, the relationship between family caregivers and paid home health aides and attendants is an important dimension of care. The relationship can be cooperative, combative, or impersonal. In general, home care workers are at the low end of the income scale, whether they are paid by agencies or directly by the patient or family; many work part-time. They often lack benefits such as health care insurance. Many are inadequately trained and supervised. Opportunities for advancement are few, and turnover is rapid. The largest segment of the workforce is made up of poor, middle-aged minority women; recent immigrants, who often lack English language skills, are another source of paid assistance for family.[14] Agency home care workers who find a family-paid job in which they are comfortable often prefer this situation to a job where they are reassigned frequently, have no role in determining how to do their work, and are subject to bureaucratic constraints. For the purposes of this essay, the main point is that the affordability of a range of services on the private market, for which the fee structure is set at the rate approved by public payors, affects the ability of families to pay for outside help and sustain long-term home care.

The physical and emotional stresses and economic strains that accompany a family's struggles to provide a sufficient level of care solely or largely on their own affects the intimate family relationships that are generally thought of as beyond the public realm. According to Noddings, "It is false to split public and private life by splitting practical from emotional and ethical matters. Community and government cannot,

---

[14] Feldman 1997.

perhaps, provide love and affection, but they can provide respect and concern."[15]

The impact of the public sector goes beyond the individual home care a family might purchase to supplement their caregiving. The availability of Medicare or Medicaid funding for a community-based program, such as adult day care, will affect the pricing schedule for the private payers. A decrease in this funding will generally mean an increase in private rates. Some elderly people can pay for services on their own, but many, especially elderly women, are poor. A daughter who is able to keep her job and support her children, while also caring for her elderly mother at home, may depend on a low fee for adult day care. If she loses this option, she is then faced with the choice of leaving her job (a loss not only to her and her family but also to her company and the economy) or paying the extra money and depriving one of her children of some important service or placing her mother in institutional care. These so-called "private" choices have been triggered by public policy. Moreover, another link between the public and private sectors is the variety of state programs under which family members are paid to provide the care to their relatives.[16]

The way in which middle-class or wealthy families undertake or fail to undertake the private responsibility of caring for ill or elderly relatives is influenced not just by public policies but also by the relative's own savings or resources, policies of the private economic and health care sectors—insurance plans, managed care organizations in their various formats, the labor market, tax structures, and so on. An employer's policies toward flextime, family leave, and emergency days off, as well as the type and limitations of the health insurance plan (if any) provided through the job are often decisive in whether the family is able to provide care at home for an ill relative.

As a part of exercising its social responsibility on behalf of all citizens, the public sector should consider ways to make it feasible—either directly through public policies or indirectly through economic incentives or regulation—for working poor or middle-class spouses to provide home care without having to become impoverished to be eligible for Medicaid. The public (as a community of taxpayers) will benefit by keeping people off subsidized programs; the public (as a community with some shared values) will benefit by supporting families' desire to provide care.

The professional sector is just as essential. To date, physicians have not played a major role in home care, even though they are accountable, but not reimbursed, for the care provided. Medical education typically does not include home care in its curriculum. A

---

[15] Noddings 1984.
[16] Linsk, Keigher, Simon-Rusinowitz, and England 1992.

survey of 123 medical schools found that only 15 schools required students to have a home care experience (some programs require only a single home visit) in the first two years, and 27 more required it in the last two years. The same survey found that in only three U.S. medical schools do all students make six or more home visits in the clinical years of training.[17]

The medical model for organizing services for elderly or disabled people is often criticized because of the tendency of some practitioners to look at disease apart from the rest of the individual's life. Nevertheless, medical care is an increasingly important part of home care, particularly as high-technology medicine is available and aggressively marketed for home use. The old medical model will have to give way to a new model that deals with the whole patient, indeed the whole family. Care plans will have to provide for sufficient medically trained professionals and paraprofessionals as well as for adequate training of family caregivers.

Because of its aggressive pursuit of higher profits through cost-cutting, achieved partly by decreased or denied hospital stays, managed care is one of the most significant sources of increased pressure on families. There has been little evidence so far that managed care takes family caregivers into account in its policies and practices. Nevertheless, managed care offers an opportunity to take a more holistic view, especially if the managed care organization has contractual obligations to serve the family. Preventing harm to family caregivers through unrealistic expectations should be part of a home care plan. Even if the managed care organization does not have obligations to other family members, it is prudent to construct a plan that will decrease the likelihood of poor quality home care or caregiver exhaustion leading to hospitalizations and more costly care.

Present "public/private" distinctions force many families into an all-or-nothing caregiving situation. Policies should be more flexible so that a wide range of families can afford a reasonable level of home care services to supplement their own caregiving and to provide services that require professional training. There is understandably great reluctance to open even the slightest possibility of expanding rather than contracting the health care economy, especially the fastest-growing segments such as Medicare home care, which has proved so profitable to so many entrepreneurs. More rigorous monitoring and fair capitation rates are justified in this area as in other sectors of the health care economy. Nonetheless, costs should not be shifted unfairly to families, who are at risk for loss of employment, housing, educational and career opportunities for other members of the family; poor health and increased health care costs for caregivers; and dissension within the family leading

---

[17] Steel, Musliner, and Boling 1994.

at times to disruption of family ties. To avoid these devastating social and human costs, public and private agencies should develop creative strategies to support family caregiving in tangible and equitable ways.

III. Differences in family capacity for caregiving

One of the major goals for the future of home care should be to recognize and respond appropriately to many kinds of diversity. Some of these are diversity of family structures and relationships; sources of strength and stress; caregiving needs required for particular medical, behavioral and social conditions; duration and intensity of caregiving; and the availability of formal and informal resources. Differences in family circumstances are crucial factors in determining the level and kind of family responsibility that is both fair and achievable. Some families cope very well in exceptionally trying situations with little or no outside assistance. Some may provide an excessive level of control, limiting the independence and autonomy of their relative. Others find the burdens overwhelming, resulting in substandard home care, potential abuse and neglect, and eventual institutionalization for the relative, poor health for the caregiver, and sometimes irrevocable family dissension and impoverishment.

Current policies and programs generally make only rough assessments of these crucial differences. Discharge planning focuses on the immediate, short-term care plan; what can be done for a few weeks or months may not be supportable for longer periods. The patient's status as Medicaid- or Medicare-eligible may be the starting point for a plan, rather than a resource to be used as appropriate. While the development of particular measures is beyond the scope of this essay and the expertise of its author, such efforts would be helpful in making a comprehensive assessment of family strengths and limitations and matching the results with individual care plans. They could be used as well in developing systems that can respond more flexibly to meet a wide range of needs.

It is important to weigh family capacity as well as patient need in determining an appropriate plan. "Capacity" is shorthand for the many factors that affect a family's ability to provide and sustain long-term home care. In a fully developed measure there would be more precise ways to rank and weigh various factors that together make up a determination of capacity. For this introductory purpose it is assumed that high, moderate, and low degrees of capacity can be assessed. Among the tangible factors are the number of people available to provide care, family economic resources, the home setting (whether the caregiver lives with the patient or not), distance of other family members from the patient's home, other family responsibilities such as small children, and other ill or elderly members. Also important are intangibles such as family dynamics; whether the family is "individualistic" (members tend to maintain separate albeit loving relationships) or "collectivist" (members tend to see the family as a cohesive unit); and cultural and religious traditions. "Social support" should also be taken into consideration. For example, formal memberships in religious or service organizations, and informal relationships with congregation members or others, are also significant in the calculus.

Many families do not fit clearly into the "high," "moderate," or "low" capacity category. Some of their characteristics fit one category, but other features fall into another group. A family may, for example, have many available members but a poor ability to cooperate among themselves. Or they may have substantial financial resources but also many members with compelling medical, educational, or other needs. Situations in which there is a primary caregiver but little family or social support are probably among the most vulnerable.

Patient needs include not only strictly medical needs but also assistance with daily activities, paying bills and managing other financial matters, emotional support, and social services. These too may not be consistent. A patient with high needs in one area may have only moderate or low needs in another area. For example, a person with a spinal cord injury may require considerable assistance in dressing and other activities of daily living but may be totally independent in managing financial matters and other intellectual activities. On the other hand, a person with a progressive disease that affects mental functioning may require minimal physical assistance and extensive behavioral monitoring.

In some cases patient needs are matched by family capacity. A patient who needs extensive care of all kinds may be fortunate enough to have family members and friends in the community who can meet all those needs without excessive drain on their personal and familial resources. Or a patient who needs relatively little assistance can be quite adequately cared for by a family with modest resources. But for most situations the match is imperfect. Patients may need help of a kind and intensity that a family is hard put to provide. Family members may be devoted and caring but simply live too far away to provide ongoing daily care. Or they may be able to manage the nontechnical aspects of care but not a highly complicated medical regimen.

The intersection of family capacity and patient needs defines a range of services and supports. Some of these can be provided by the family, some by community or volunteer resources, and others by formal, professionally trained personnel. This discussion excludes situations in which family members refuse, for whatever reason, to accept any significant responsibility for caregiving, and situations in which there is no family or the family is so incapacitated by other problems that it cannot provide care. Thus the discussion focuses on the varieties within the middle ground—families in which there is a willingness to provide care but varying levels of capacity and need.

To illustrate how such a measure might help families and care planners arrive at a reasonable set of services, consider the following examples. Patient 1 is a 77-year-old man with colon cancer. His first bout of cancer occurred ten years ago, and he has been treated for recurrences in the past three years. He has become increasingly dependent. He is being treated with IV home chemotherapy and requires frequent medical monitoring and adjustment of medications. He is weak and needs assistance in Activities of Daily Living (ADLs), such as bathing, toileting, feeding, dressing, and moving from bed to chair. He can only eat specially prepared food. Although there is some hope that the treatment will lead to remission, the oncologist is beginning to talk about palliative care and perhaps hospice. Family 1 consists of his wife of forty years, one married daughter, one married son, and one unmarried son. The patient also has a sister and brother whose families have been very close to him. He is a prominent member of his church and has been active in several civic groups. He has strong ties in his community. The whole family has discussed the division of labor at key points in the illness; though the tasks are not divided equally, there is family unity about what is needed and how it should be provided. Through the illness the patient's wife and his daughter have provided most of the care, with frequent help from the other children and his sister's family. Although many of his medical costs are not covered by Medicare or his supplemental insurance, the unmarried son, who has a thriving business in another city, has made significant financial contributions. Family 1 is understandably upset at the evident decline of their relative; they need counseling about the options available for both family and patient and dealing with the anticipation of his death. They also need home care visits by trained medical professionals and instruction in how to manage the care regimen and potential emergencies. They need someone to answer their questions. The members of the family can relieve each other and provide the needed respite. Community members can help with food, shopping, chores, and emotional support as well. This family is going through a crisis; they need help in managing it, but they have the necessary family strength and resources to handle the challenge.

Family 2, assuming the same type of condition but a patient without the same level of community involvement, has much greater needs for formal help. This man is divorced; his married son has been estranged for years; his daughter, the child closest to him emotionally, lives the furthest distance from him. She visits a few times a month and takes over complete care while she is there. The unmarried son (not so well-to-do as in Family 1) is able to provide some care, particularly in the evenings and on weekends. There are fewer material resources with which to hire formal caregivers; much more subsidized assistance is necessary to keep him at home through most of his illness.

Family 3, again with the same type of patient, has even fewer resources. All three children live in distant cities; one is in the armed forces, one has a severely handicapped child, and the third is struggling to overcome alcoholism. Only an elderly sister and a niece live nearby; neither speaks English fluently and both are easily intimidated by the bureaucratic systems they encounter in trying to obtain help. They try to do what they can, but they are soon overwhelmed by the intensity of the patient's needs. Physicians, nurses, and home health aides are needed to stabilize and monitor the patient's condition. The sister and niece need counseling and support from the rest of the family and from their friends and neighbors to maintain their level of commitment.

These scenarios are only an approximation of the complexities of real cases and real families. They illustrate the enormous diversity of family capacities that should be considered in developing appropriate care plans. Equally important is the need to re-evaluate the care plans at regular intervals and especially in case of changes in the patient's condition, family circumstances, or other events which relevantly alter the circumstances.

These examples say nothing about an important question: Who should pay for this care? It would be too simplistic, to say nothing of unrealistic, to assume that public programs should pick up all the unmet needs for family support, once they are identified. But it would be equally simplistic to assume that families can pay for all the services they need on their own. Some combination of enhanced private insurance benefits, the services of voluntary health organizations and community agencies where appropriate, and limited public funds for the most extreme circumstances could make a major difference. Coordination of existing services and communication among service providers as well as with patients and families would in itself be a significant benefit.

Public and private funds could also be used for training professionals and paraprofessionals to recognize the special needs of family caregivers and to provide more effective services than now exist. Private foundations could also play a role in supporting and evaluating innovative programs. Most interventions with families come in times of crisis, when the fragile support systems fail and families can no longer cope. Planning at the outset that realistically assesses what families can do on their own, and where they need support, and then fills the gap to the extent possible, should in the long run prove not only more compassionate but also more effective. Moran asserts that "instead of urging people to accept their responsibilities, we would do better to try to develop their capacity for response."[18] This holds true for professionals and public and private agencies that serve patients and families. They too must develop a stronger capacity for response.

---

[18] Moran 1997, p. 82.

IV. Summary and recommendations
To summarize, this essay makes several key points:

1. Home care in the future will have to recognize and respond appropriately to a greater diversity of families and more medically complex patient needs.

2. A new model of care, integrating medical and social concerns, should be developed so that the patient, family, and professionals are addressed as a complex and dynamic partnership.[19]

3. As a concept, family responsibility does not easily fit with more legalistic, contractual notions of responsibility. It is an amorphous concept, with fluid boundaries and interpretations. It may derive from religious belief, cultural tradition, emotional bonds, gratitude for past acts, or a sense of obligation apart from love.

4. There are limits to what can reasonably be expected of families, although the boundaries will vary considerably in each case.

5. The sectors that are variously labeled public, private, corporate, professional, and governmental have distinctive characteristics but they are also interrelated. Decisions made in each sector profoundly affect the others.

6. Family capacity—the strengths and limitations that affect a family's ability to provide care, particularly over the long term—should be assessed at the outset of providing care and periodically thereafter.

7. Society has a responsibility to support families that take on the exceptional responsibilities of providing care, even at considerable cost to themselves.

Recommendations
These recommendations are intended to be mid-level recommendations, that is, not so grandiose that they depend on vast changes in the American political or economic scene nor so trivial that

---

[19] Levine and Zuckerman 1998.

they accomplish little more than a token bow to family caregivers. These are recommendations that can be achieved.

1. The technical, psychosocial, and financial aspects of home care should be part of the curriculum of medical and social service professionals. This training should include an understanding of the contribution and stresses of family caregivers and, at a minimum, several home visits. Physicians, nurses, and social workers should have the experience of directly providing care in the home, so that they learn first-hand how a home and a hospital differ. Family caregivers should be used as teachers as well as learners; they have the direct knowledge of how home care actually takes place.

2. Model programs should be developed in home care agencies, hospitals, rehabilitation centers, or community-based agencies that offer innovative ways of involving families in decision making, training, and assessing family capacity at various points in the course of care. Transition points in the course of illness or care settings is one potential focus for such programs.[20]

3. Family advisory councils should be created or given enhanced roles in home care agencies and managed care organizations to provide feedback on service implementation, proposed policies, and other issues.

4. Public policy leaders should explicitly consider the "family impact" of decisions on home care policies and programs in terms of access, resources, and support.

5. Community voluntary agencies and religious organizations should consider how they can supplement family caregiving with volunteers. The goals should be to provide assistance with the family's own assessment of needs, not to substitute for formal care.

6. A thorough review of the literature and practice of respite programs should be conducted to identify key elements in successful and unsuccessful programs.

---

[20] Levine 1998.

An agenda for the future must include the people who provide most of the nation's health care (families) and the setting in which most care is actually provided (the home). To paraphrase Freud, "What do family caregivers really want?" For some the answer may be, "A good night's sleep." For others, better communication with professionals is essential: "Someone to call who can answer my questions." For still others, "More help from the rest of the family." For some, better insurance coverage or assistance with formal home care will make the difference; for others, more room in the household. Some may want nothing. And for some, none of these measures is adequate. Only removing the burden of responsibility will suffice. All these views and others are part of family caregiving; they deserve respect and attention.

References

Abel, E.K. 1995. "'Man, Woman, and Chore Boy': Transformations in the Antagonistic Demands of Work and Care on Women in the Nineteenth and Twentieth Centuries." *The Milbank Quarterly* 73(2): 187-211.

Bendetson, J. 1997. "I Am More Than Hands." *New York Times Magazine* (April 13).

Covinsky, K.E., Goldman, L., Cook, F., Oye, R., Desbiens, N., Reding, D., Fulkerson, W., Connors, A., Lynn, J., and Phillips, R.S., 1994. "The Impact of Serious Illness on Families." *JAMA* 272(23) (December 2): 1839-1844.

Dubler, N.N. 1990. "Accommodating the Home Care Client: A Look at Rights and Interests." In Home Health Care Options: A Guide for Older Persons and Concerned Families, eds. C. Zuckerman, N.N. Dubler, and B. Collopy. New York and London: Plenum Press: 41-166.

Feldman, P.H. 1997. "Labor Market Issues in Home Care." In Home-Based Care for a New Century, eds. D.M. Fox and C. Raphael. Malden, MA: Blackwell Publishers, Inc.

Hokenstad, A., Ramirez, M., Haslanger, K., and Finneran, K., 1997. Medicaid Home Care Services in New York City: Demographics, Health Conditions, and Impairment Levels of New York City's Medicaid Home Care Population. New York: United Hospital Fund (March).

Holstein, M. and T.R. Cole. 1995. "Long-Term Care: A Historical Reflection." In Long-Term Care Decisions: Ethical and Conceptual Dimensions, eds. L.B. McCullough and N.L. Wilson. Baltimore and London: The Johns Hopkins University Press: 15-34.

Institute for Family Centered Care. 1994. "Reinventing Family Policy." *Advances* 1(1) (Spring).

Institute for Health and Aging at the University of California, San Francisco. 1996. "Chronic Care in America: A 21st Century Challenge." Princeton, NJ: The Robert Wood Johnson Foundation.

Jonsen, A.R. 1968. Responsibility in Modern Religious Ethics: A Guide to Making Your Own Decisions. Washington, D.C.: Corpus Books.

Keigher, S.M. and R.I. Stone. 1994. "Family Care in America: Evolution and Evaluation." *Aging International* (March): 41-48.

Levine, C. 1991. "AIDS and Changing Concepts of Family." In A Disease of Society: Cultural and Institutional Responses to AIDS, eds. D. Nelkin, D.P. Willis, and S.V. Parris. Cambridge: Cambridge University Press: 45-70.

Levine, C. 1998. "Rough Crossings: Family Caregivers' Odysseys through the Health Care System." New York: United Hospital Fund.

Levine, C. and C. Zuckerman. 1998. "The Trouble with Families: Toward an Ethic of Accommodation." Annals of Internal Medicine (Forthcoming).

Linsk, N.L., Keigher, S.M., Simon-Rusinowitz, L., and S.E England. 1992. Wages for Caring: Compensating Family Care of the Elderly. New York: Praeger.

Manton, K.G., L.S. Corder, and E. Stallard 1993. "Estimates of Change in Chronic Disability and Institutional Incidence and Prevalence Rates in the U.S. Elderly Population from the 1982, 1984, and 1989 National Long-Term Care Survey." *The Journals of Gerontology* 48, no. 4: S153-S166 (July).

Marks, N. F. 1996. "Caregiving Across the Lifespan: National Prevalence and Predictors." *Family Relations* 45: 27-36 (January).

Moran, G. 1996. A Grammar of Responsibility. New York: The Crossroad Publishing Company.

National Alliance for Caregiving. 1997. Family Caregiving in the U.S.: Findings from a National Survey (June).

Nelson, H.L. and J.L. Nelson. 1995. The Patient in the Family: An Ethics of Medicine and Families. New York: Routledge.

Noddings, N. 1984. "The Cared-For." *In Caregiving: Reading in Knowledge, Practice, Ethics, and Politics*, eds. S. Gordon, P. Benner, and N. Noddings. Philadelphia: University of Pennsylvania Press: 21-39.

Nottingham, J.A. et al. 1993. "Caregivers and Caregiving in West Central Georgia: Highlights of the Care-Net Study." A report of the Rosalynn Carter Institute of Georgia Southwestern College: Americus, GA (July).

Post, S.G. 1995. The Moral Challenge of Alzheimer Disease. Baltimore and London: The Johns Hopkins University Press.

Sevick, M.A. 1996. "Economic Cost of Home-Based Care for Ventilator-Assisted Individuals." *Chest* 109(6) (June): 1597-1606.

Smith, P. 1993. "Family Responsibility and the Nature of Obligation." In Kindred Matters: Rethinking the Philosophy of the Family, eds. D.T. Meyers, K. Kipnis, and C.F. Murphy, Jr. Ithaca and London: Cornell University Press: 41-58.

Steel, R. K., M. Musliner, and P. A. Boling. 1994. "Medical Schools and Home Care (letter to the editor)." *New England Journal of Medicine* 331, no. 16: 1098-1099 (October 20).

Acknowledgement

This essay is derived from a concept paper prepared for the Visiting Nurse Service – Robert Wood Johnson Home Care Research Initiative; the permission of both organizations to adapt the paper for this volume is gratefully acknowledged. Many people contributed ideas, personal and professional experiences, and specific suggestions to the development of this paper. These included members of the group convened by the Visiting Nurse – Robert Wood Johnson Home Care Research Initiative, the Advisory Committee of the Families and Health Care Project of the United Hospital Fund, and the participants in two meetings on family responsibility organized by the Families and Health Care Project. Penny Hollander Feldman, Kenneth Covinsky, R. Knight Steel, and two anonymous reviewers offered valuable comments on the manuscript at various stages. Finally, colleagues at the United Hospital Fund contributed significantly in editorial, research, and other areas to the development of the manuscript. All these individuals made valuable suggestions; in the spirit of this paper, I accept responsibility for the final product.

ETHICS AND PUBLIC POLICY IN A DEMOCRACY:
THE CASE OF HUMAN EMBRYO RESEARCH
William A. Galston

A brief political history

For more than a quarter of a century, the issue of research on human embryos has been entangled in broader controversies over fetal research and abortion. During the late 1960s and early 1970s, policy makers at the National Institutes of Health (NIH) convened a study panel to consider regulations covering fetal research. Publication of draft fetal research guidelines aroused little interest until the *Roe v. Wade* decision in 1973.

Soon afterwards, the issue moved from the bureaucratic arena (what one scholar calls the "science subgovernment") to the public domain, where it became a "small skirmish in the larger battle over abortion."[1] The *Washington Post* published a front-page article entitled "Live-Fetus Research Debated," which sparked national news coverage and a demonstration at NIH headquarters. In 1974, Congress suspended all federal support for nontherapeutic fetal research until a national commission developed appropriate ethical guidelines. Within a year, the commission had made its recommendations, and Secretary of Health and Human Services Caspar Weinberger was able to sign a set of regulations creating a federal ethics advisory board (EAB) whose review and advice would be needed to fund most categories of fetal research.

Among other matters, the EAB had jurisdiction over in vitro fertilization (IVF), which it considered especially between 1978 and 1980. In one report issued during that period, the EAB concluded that federal funding of IVF research, subject to certain safeguards, was ethically appropriate. The HHS Secretary took no action on that report, and the charter of the EAB lapsed in 1980, shortly before the election of Ronald Reagan. Repeated efforts between 1981 and 1993 to reconstitute the EAB all failed, and in the ensuing absence of EAB review a "de facto moratorium" on federal funding of fetal research prevailed.

The election of Bill Clinton created a new situation. Within days of assuming the presidency in January 1993 he issued an executive order lifting the ban on fetal tissue research. In June, he signed into law the NIH Revitalization Act of 1993, which nullified the regulation requiring EAB review of embryo research proposals. The NIH had already received a number of proposals for research on "preimplantation

---

[1] Steven Maynard-Moody, "Managing Controversies over Science: The Case of Fetal Research," *Journal of Public Administration Research and Theory* 5 (1995): 10.

embryos" (in vitro embryos in early stages of development prior to implantation in the uterus) and soon received more. Rather than rushing to fund these proposals, however, NIH Director Harold Varmus established an external panel to review the ethical issues raised by the use of human preimplantation embryos in research and the recommended guidelines for the funding and conduct of such research.

The Human Embryo Research Panel, whose nineteen members included philosophers, theologians and legal scholars as well as research scientists, met five times between February and August of 1994. Early on, there were indications that its proceedings would be carefully monitored and that its recommendations would prove politically controversial. On June 16, 1994, Varmus received a letter signed by thirty-five members of Congress posing a series of legal and ethical questions concerning the Panel's work. The tone and content of these questions made it clear that the letter was intended as a warning-shot.[2] Just five days later, Varmus responded with a lengthy description of the Panel's purposes and procedures. He defended the Panel as an appropriate means to carry out the intent of Congress in the NIH Revitalization Act of 1993 and assured the legislators that the Panel was carefully considering the kinds of ethical issues that troubled them.[3] Later, the members of Congress responded to Varmus with another letter stating that in their view, his letter raised more questions than it answered. They were particularly troubled by the Panel's apparent failure to include embryos within the purview of human subjects entitled to a wide range of protections during the conduct of research.[4]

The Human Embryo Research Panel released its report on September 27, 1994, sparking the events that constitute the primary focus of this essay.

The report and its aftermath

The Panel's thoughtful and detailed report dealt with issues ranging from scientific, medical and ethical issues in preimplantation embryo research to guidelines for such research and specific categories of activities that should qualify for (or be denied) federal funding. The Report's recommendations were based on three principal considerations:

- The promise of human benefit from research is significant, carrying great potential benefit to infertile couples, and to families with genetic conditions, and to individuals and

---

[2] Letter of Rep. Robert K. Dornan, et al. to Dr. Harold Varmus, June 16, 1994.
[3] Letter of Harold Varmus to Rep. Robert K. Dornan, June 21, 1994.
[4] Letter of Rep. Robert K. Dornan, et al. to Dr. Harold Varmus, September 19, 1994.

families in need of effective therapies for a variety of diseases.

- Although the preimplantation human embryo warrants serious moral consideration as a developing form of human life, it does not have the same moral status as infants and children. This is because of the absence of developmental individuation in the preimplantation embryo, the lack of even the possibility of sentience and most other qualities considered relevant to the moral status of persons, and the very high rate of natural mortality at this stage.

- In the continued absence of Federal funding and regulation in this area, preimplantation human embryo research which has been and is being conducted without Federal funding and regulation would continue, without consistent ethical and scientific review. It is in the public interest that the availability of Federal funding and regulation should provide consistent ethical and scientific review for this area of research.[5]

The question of ethically acceptable sources of preimplantation embryos was, as the Report makes clear, one of the most difficult issues the Panel had to consider. There was a broad consensus among Panel members that embryos remaining from IVF treatments and donated by women or couples (the so-called "spares") could be used for research. But the Panel wrestled with the issue of whether it was ethically permissible to create embryos expressly for research purposes, without the intent or expectation that they would ever be implanted and develop into infants. The Panel believed that it should approach this issue by balancing the health and safety of men, women and children against the moral respect due the preimplantation embryo. It concluded that while the needs of men, women and children should be given priority, the moral status of the preimplantation embryos should limit their use to the most compelling circumstances: when the research cannot otherwise be conducted, or when it is necessary for the validity or statistical power of a study that is potentially of outstanding scientific and therapeutic value. In no case should research be conducted on embryos beyond the fourteenth day of development, at which point irreversible individuation typically occurs.[6]

---

[5] Final Report of the Human Embryo Research Panel, National Institutes of Health, September 27, 1994, p.2.

[6] Report, p.4.

This recommendation did not enjoy unanimous support. Patricia King, a professor of law and one of the Panel's co-chairs, dissented in part:

> The prospect that humanity might assume control of life creation is unsettling and provokes great anxiety. The fertilization of human oocytes for research purposes is unnerving because human life is being created solely for human use. I do not believe that this society has developed the conceptual frameworks necessary to guide us down this slope.... At the very least, we should proceed with extreme caution. Perhaps the public's concerns can be allayed over time with the development of appropriate conceptual frameworks. In any event, the public must be convinced that such actions are necessary to obtain significant benefits for human kind and that the research will be responsibly conducted.[7]

King was among the first to raise an issue that became central in the next few months—the role of public opinion in a democracy. In a benevolent dictatorship, the claim that a proposed course of action will promote the public interest would (if true) suffice to legitimate that course of action. But in a democracy, there is no class of Platonic guardians. The people must decide for themselves whether to accept or reject proposed public acts. Scientific and medical experts may seek to shape and inform public judgment, but they cannot substitute themselves for it. The fact that the public finds a proposed course of action ethically "unsettling" or "unnerving" is a matter of legitimate consequence in a democracy—even if the public cannot articulate doubts with the kind of conceptual clarity demanded by moral philosophers.

Within a few days it became clear that King was not alone in her doubts and that a firestorm was developing. The disapproval of anti-abortion activists was to be expected. What was not expected was the forceful intervention of the mainstream media, spearheaded by the *Washington Post*. In a lead editorial, the *Post* blasted the Panel's recommendation:

> The creation of human embryos specifically for research that will destroy them is unconscionable. The government has no business funding it.... Is there a line that should not be crossed even for scientific or other gain,

---

[7] Id., p.97.

and if so where is it?.... In approving
the funding of the purposeful creation of
human embryos for any experiments the
panel took a step too far.[8]

The *Post*'s editorial was swiftly followed by a torrent of editorials, op-ed
pieces, and letters to the editor in newspapers around the country.
Meanwhile, alarm-bells were going off in the White House, where I was
then serving as Deputy Assistant to President Clinton for Domestic
Policy. The issue had clearly moved out of the inner workings of the
bureaucracy into the public arena. The President would be held
accountable for the actions of the executive branch.

To ensure that the issue would be carefully monitored and that
the President would receive informed advice, and ad hoc working group
was established, led by Chief of Staff Leon Panetta, Deputy Chief of
Staff Harold Ickes, and Senior Advisor for Policy George
Stephanopoulos. The working group included representatives of the Vice
President's office, the First Lady's office, the Counsel's office, the
Office of Science and Technology Policy (OSTP), and the Domestic
Policy Council (DPC), among others.[9]

The day after the publication of the Panel's report, John
Gibbons, the Assistant to the President for Science and Technology,
wrote a memo to Panetta summarizing the procedures NIH would
employ to review the report before reaching final recommendations.
(Throughout the fall, the OSTP vigorously represented the views of the
scientific community.) In response to a request from the Deputy Chief of
Staff, and drawing especially on moral and political philosophy, I
summarized the ethical issues raised by the report. Joel Klein, the Deputy
Counsel, took the lead on legal issues. On November 9, 1994, NIH
Director Harold Varmus came to the White House to give the working
group a detailed briefing on the scientific issues. On the 21st, Klein and
Gibbons wrote a memo to Panetta analyzing the advantages and
disadvantages of three options: full support for the Panel's
recommendation; rejection of its recommendation concerning embryos
created expressly for research purposes; and an outright ban on all
federal funding for human embryo research. The full working group met
several times, and members conferred informally on a regular basis.

By late November, we had reached two conclusions: first, that it
would be very difficult to offer a compelling public justification for the
Panel's recommendation that the federal government should fund the
creation of embryos specifically for research purposes; and second, that

---

[8] *Washington Post*, October 2, 1994, C6.
[9] I served as the lead DPC representative.

the matter had to be resolved decisively. We were working against a deadline: the Advisory Council to the NIH Director was meeting in early December to consider the Panel's recommendations. We expected the Advisory Council to ratify those recommendations and advise the Director to accept them, which we assumed he would—perhaps the same day. So we recommended to the President that he personally intervene to prohibit the NIH from funding the creation of human embryos specifically for research purposes.

The President accepted our advice. In a written order released December 2, 1994 (the day of the Advisory Council's meeting), he acknowledged his appreciation for the work of the NIH committees and recognized the important scientific and medical benefits that could flow from research on human embryos. Nevertheless, citing "profound ethical and moral questions as well as issues concerning the appropriate allocation of federal funds," the President declared: "I do not believe that Federal funds should be used to support the creation of human embryos for research purposes, and I have directed that the NIH not allocate any resources for such research."[10]

In response to numerous press inquiries, members of the President's staff made it clear that this order referred only to embryos created solely and specifically for research purposes. The President had no intention of challenging the scientific consensus in favor of funding research of embryos created in the course of normal IVF procedures.

The President's distinction between the two categories of embryos was criticized from two very different directions. Many members of the scientific community (including the President's own science advisor) saw no distinction, particularly when vital experiments could not proceed without research embryos. If research on spares is acceptable, why not research on embryos created for that purpose?

Leaders of the Catholic Church also denied the relevance of the distinction, but they drew the opposite policy conclusion. Cardinal Keeler, the President of the National Council of Catholic Bishops, wrote that

> a hard-and-fast distinction between "spare" and "research" embryos is untenable....This distinction means that human embryos cannot be created as part of a federally funded experiment, but can be manipulated and destroyed so long as they come from outside the federal project. Such a policy— that government may not use tax dollars to create life but only to destroy it—defies all moral logic.[11]

---

[10] Statement by the President, The White House, December 2, 1994.
[11] Letter of December 7, 1994 to President Clinton.

Despite these criticisms, I remain convinced that President Clinton's position was sound and prudent. In the following section of this article, I offer a sketch of the considerations that appeared compelling to most members of the White House working group and ultimately to the President.

Practical and moral considerations in public decision-making

Deliberation never begins with a blank slate. Over the past two decades, all 50 states and many foreign countries have considered human embryo research and have passed legislation to regulate it. While every foreign country allows in vitro fertilization procedures, the vast majority impose significant restrictions on embryo research. (Of the eleven countries surveyed by the Panel, only four permitted research as extensive as the Panel proposed, and one banned it altogether.) These worldwide deliberations created a body of shared scientific, ethical, and institutional propositions from which U.S. federal policy makers were able to draw.[12]

In addition to this legislative and policy background, members of the White House working group identified what the philosopher John Rawls has called "provisional fixed points"—specific practical judgments that fall short of certainty but enjoy widespread (not necessarily unanimous) support and are entrenched in our public culture. Four such judgments seemed especially pertinent.

Our **first** "fixed point" was the distinction between abortion and contraception. While we recognized that from an ethical (as opposed to legal) standpoint the issue of abortion remains unsettled in American public culture, this is not the case for most forms of contraception, which enjoy broad public support. We were therefore unwilling to embrace any position whose logic implied the ethical rejection of contraceptive strategies such as intra-uterine devices. (In this respect, we concurred with the Panel's reasoning.)[13]

Second: we assumed that in vitro fertilization procedures (in roughly their current form) are ethically appropriate. The purpose of these procedures—enabling otherwise infertile couples to have healthy, normal children—is ethically valid, as is what is necessary to achieve this purpose, including the fertilization of more embryos than can be used for possible implantation.

Third: we distinguished between private conduct and public funding. Taking our cue from the long running controversy over federal

---

[12] For the specifics, see Lori B. Andrews, "State Regulation of Human Embryo Research," and Andrews and Nanette Elster, "Cross-cultural Analysis of Policies Regarding Human Embryo Research" (papers commissioned by the Human Embryo Research Panel).

[13] Report, p. 46.

support for abortions conducted under the aegis of Medicaid, we concluded that it was ethically coherent for the government to refrain from funding an activity that was legally permissible. It is one thing for the government to respect the liberty of individuals, but a very different matter for the government to act in ways that imply support for particular choices that individuals make in the exercise of their liberty. Government funding is not an ethically neutral act; it implies a degree of endorsement, and it implicates every taxpaying citizen in the funded activities. In circumstances of deep moral division, it is typically unwise for the government to move beyond legal acquiescence (which is itself controversial) to material support.

Finally: we assumed (as did the Panel) that it was appropriate to impose ethical limits on the conduct of scientific inquiry. We were well aware of the long historic link between the process of scientific discovery and the betterment of the human condition. We were also aware of the fact that at various points in that history, ethical and theological objections to particular scientific procedures (such as the dissection of cadavers) had been swept aside, to the long-term advantage of our species and with moral consequences retrospectively judged to be acceptable. Nonetheless, in the wake of events such as the Nazi experiments on human subjects and the Tuskegee experiment (for which President Clinton has apologized on behalf of the American people), most people believe that science cannot be treated as ethically autonomous. The premise, "Experiment X is essential for the pursuit of knowledge" is not sufficient to warrant the conclusion that "Experiment X is ethically acceptable." That is the case *even if* the knowledge to be gained is potentially of great benefit to humankind.

There are two different ways in which ethical considerations might limit scientific inquiry. One is a kind of "balancing test" in which ethical considerations are placed on the deliberative scales along with other considerations, such as the practical importance of the knowledge to be gained. This is the procedure the Panel employed in reaching its conclusion about the creation of human embryos for research purposes: the large potential benefits were held to outweigh the ethical costs of using embryos whose moral status is lower than that of actual human beings.

But there is another way in which ethical considerations can constrain scientific inquiry. These considerations can serve as "side-constraints"—that is, as absolute barriers to a particular course of action, whatever the projected benefits of that course may be. (Immanuel Kant's injunction to act so as to treat humanity as an end in itself, and never as a means alone, is one important basis for this way of thinking.) From this standpoint, the use of a balancing test is ethically inappropriate, because it falsely assumes that the ethical considerations in question can be "traded off" against others. Notably, the Panel was willing to employ a version of the side-constraint strategy, rooted in widely held moral beliefs, to address issues such as cloning:

> The notion of cloning an existing human being, or of making "carbon copies" of an existing embryo, appears repugnant to members of the public. Many members of the Panel share this view, and see no justification for Federal funding of research ... for this purpose.[14]

The question of which ethical strategy is preferable cannot be detached from the features of specific controversies. When the issue involves experimentation on human subjects without their consent, most people consider the balancing test to be wildly inappropriate. (Could any projection of scientific value have justified the Tuskegee experiments?) Matters are more difficult when, as in the dispute over embryo research, there is sharp disagreement over the moral status of the experimental subject. Nonetheless, in considering the issue of embryo creation for research, many members of the White House working group came to believe that the side-constraint strategy of ethical reasoning was more appropriate than the balancing test the Panel had used, and that public objections to embryo creation for research purposes were as intense and as plausible as were their objections to cloning.

In reaching this conclusion, we were aware of the pitfalls of relying on moral sentiments and intuitions (which some of us came to call the "yuck factor.") In the past, after all, large numbers of people expressed their repugnance for scientific procedures and social practices that are now widely accepted. In some cases, "natural" sentiments were invoked to defend practices (such as prohibitions on interracial marriage) that could not be rationally defended.

Still, officials in a democracy who act in the name of the people must begin by taking public sentiment seriously. Of course, our responsibility did not end there. If we had concluded that the people were clearly mistaken, we would have had the obligation to enter into a public dialogue with them in an effort to change their minds. But while we recognized a range of plausible views on the status of human embryos, it was not clear to us that the center of gravity of public opinion was in error. It was therefore entitled to a substantial measure of democratic respect.

Many of us were fortified in our view by a consideration that Patricia King had raised in her dissent. Our society, she suggested, has not yet developed the conceptual apparatus needed to deal with embryo research in a nuanced way. If so, then the absence of a hard-and-fast side-constraint is likely to produce a slippery slope on which more and

---

[14] Report, p. 94.

more research is considered legitimate, at least within the scientific community. To be sure, the Panel itself sought to construct one clear barrier—the fourteen day limit. But along with a number of editorial writers, we were not reassured by indications that the Panel was not fully committed to that limit. The crucial passage of its Report reads, "*For the present*, research involving human embryos should not be permitted beyond the time of the usual appearance of the primitive streak in vivo (14 days)."[15] This seemed to us symptomatic of the fact that most members of the scientific community were chafing against limits to research and that it was therefore necessary to clarify and strengthen those limits.

The White House working group disagreed with the Panel on another key issue—the nature of the dialogue appropriate for public reflection in a liberal democracy. In an effort to define a critical element of this dialogue, the social philosopher Rawls suggests the employment of scientific reasoning "when not controversial." The Panel actually cited Rawls but dropped his qualifier:

> Public policy employs reasoning that is understandable in terms that are independent of particular religious, theological, or philosophical perspective, and it requires a weighing of arguments in the light of the best available information and scientific knowledge.[16]

One may well wonder whether there really *are* any such "independent" terms. And even if there are, it is by no means clear that they define the limits of acceptable public discourse in a constitutional democracy. Along with many philosophers, President Clinton did not accept the Panel's definition. He believed (and continues to believe) in a more robust role for faith-based arguments in the public arena. From this perspective, the challenge of public dialogue is not to screen out religious and metaphysical commitments in the name of an elusive "neutral" policy Esperanto, but rather to find ways of dealing with the deep differences we inevitably (and properly) bring into the public arena.

This is in no way to denigrate the public importance of scientific arguments. Along with the Panel, the White House working group did its best to understand the scientific research process and key findings. We concluded that while science frequently delimits the range of acceptable policy choices, it rarely prescribes a specific alternative.

Consider, for example, scientific findings concerning the onset of genuine individuation in embryo development. For several days after fertilization, embryonic cells are undifferentiated and capable of

---

[15] Report, p. 3; emphasis added.
[16] Report, p. 50-51.

developing in a range of directions. Moreover, these cells do not form part of an organized whole; one or more of them can be removed without affecting the later development of the fetus. Up to about fourteen days after fertilization, a single embryo can split into twins (or higher-order multiple births), a capacity that ends only with the appearance of the "primitive streak" that definitively establishes the embryo's head/tail and left/right orientation.[17]

So far, so good: but what policy conclusions follow from these facts? Consider what might seem a far-fetched analogy. Suppose you are told that either a single infant or identical twins have been locked in a closet. Would you be justified in conducting an experiment (say, on the composition of the air in the room) that could jeopardize the well-being of whoever is behind the closed door? What is the moral import of your ignorance of the exact number of individuals at risk? Or suppose that you are assured that a particular experiment will have no adverse effect on the development of the infant(s). Does it follow that you are at liberty to conduct a different experiment that would have such an effect? At best, the issue of individuation is embedded in the broader question of the moral status of the developing embryo. But this question (unlike individuation) cannot be answered in purely   scientific terms.

Lurking behind these conceptual questions was an eminently political issue: when judgments concerning scientific research are contested, who should decide? Many scientists believe that these conflicts should be resolved within the scientific community itself. This proposition is exposed to two objections. First, as we have seen, to the extent that the conflicts revolve around more than science, they cannot be fully resolved through scientific procedures and arguments. And second, to the extent that the issue is not the liberty of scientific inquiry but rather (as in embryo research) the disposition of public funds, the public has a legitimate interest in the decisions made by the elected and appointed officials.

Once an issue is subject to political determination, new considerations come into play. For example, it becomes important to ask whether a position can be effectively explained to the public. In a democracy, if you can't frame good solid public arguments for your position, you have good reason not to proceed. As the chairman of the Panel, Steven Muller, himself acknowledged, "By a huge majority, the public has no idea what ex utero or pre-implantation human embryo research means or what it involves. But it does, to most people, sound terrible."[18] For this reason, among others, the White House working

---

[17] See Report, pp. 20-22 and 107; also Peter Singer, et al., Ed., Embryo Experimentation (Cambridge: Cambridge University Press, 1990), Chapters 1, 5 and 6.
[18] Quoted in *Christianity Today*, January 9, 1995, p. 38.

group feared a political backlash: if we accepted the Panel's full recommendation, adverse congressional and public reaction might well lead to the continued cutoff of all funding for embryo research. We didn't want this to happen, and we didn't see how the aims of the scientific community would be served by this outcome.

The White House working group was acting in a context of deep moral divisions, within which a neutral language was unavailable and full consensus unachievable. There was, we thought, a moral imperative to seek a defensible compromise based on the ethical center of gravity of the arguments, the long-term best interests of the country, and the common sense of the people.

The outcome: The center did not hold

In the end, the White House working group's effort to locate a viable middle ground did not succeed. As part of a temporary spending measure, the President was compelled to accept a rider banning all federal funding for embryo research.

Why did our efforts fail? The proximate cause was surely the November 1994 congressional elections, which brought to power a Republican majority strongly backed by Christian conservatives. Not surprisingly, this new majority was not disposed toward compromise and viewed the embryo research issue through the prism of a highly polarized abortion debate.

The Clinton administration also contributed to this result, however. The administration came to power in 1993 determined to reverse the restrictive abortion policies of the previous twelve years, a goal fervently advocated by core Democratic constituencies. Especially during 1993 and 1994, the administration was not in a mood to go slowly on this issue. Its first acts in January of 1993 included pro-choice executive orders, and it pushed the NIH Revitalization Act of 1993 in part to nullify research restraints seen as stemming from pro-life pressures. So conservatives had some reason to doubt that the administration was really interested in serving as an honest broker or in seeking honorable compromise on embryo research issues.

Ultimately, the failure of the center to hold was rooted in a basic structural feature of contemporary American politics. When an issue becomes visible enough to engage the attention of a mass public, the moderation and common sense of the people as a whole will typically favor middle-ground approaches. On the other hand, when an issue is contested principally among political elites and organized interest groups, positions at the far ends of the spectrum tend to predominate. (This is the case because during the past generation, the political parties have become more polarized and single-issue groups with intense points of view have proliferated.)

Despite numerous news stories and editorials during September and October 1994, human embryo research did not engage the attention of the electorate as a whole and did not function as a voting issue in November of that year. Not surprisingly, the issue subsequently became the subject of a tug-of-war between the executive branch and the congressional majority, with most of the American people firmly on the sideline. This proved to be a formula for winner-take-all politics rather than compromise.

Conclusion: The moral status of the preimplantation embryo

So much for the politics of the core issue raised by the President's order—the distinction between "spare" and "research" embryos. What about its merits? In a thoughtful review of the issues published in the Catholic journal *Commonweal*, Susan Ellis and Gordon Marino commented as follows:

> It is hard to imagine that the clock will be turned back on IVF. Consequently, a large number of embryos, which can either be discarded or used for research purposes, and then discarded, will continue to be produced. It is often argued with some cogency that in light of the substantial benefits that could come from doing research on these never-to-be-implanted embryos, it is nothing less than a sin not to use them for research purposes. If, however, we are willing to do research on embryos solely because they are no longer intended for implantation, then we are in effect stating that the moral status of the embryo does not depend upon its intrinsic properties, but rather upon the intention that others have for it…. If it is morally [appropriate] to experiment upon embryos no longer intended for implantation, then there must not be any morally compelling reason to protect embryos, period…. If one cluster of embryos is not entitled to the rights and protections of personhood, why should another be?[19]

These are powerful arguments, but again they prove too polarized for the public. Intention *is* relevant in moral considerations: in recent polls of the general populace, roughly 80 percent of Americans approve of abortion in cases of rape, versus only 20 percent who approve of it for sex selection of children. Put more generally: about one-fifth of the population disapprove of abortion under virtually all circumstances; about one-fifth approve under virtually all circumstances; about three-fifths believe that the circumstances of conception and the intention of the agent make a crucial difference. It is not necessarily a sign of moral

---

[19] *Commonweal*, December 2, 1994, p. 9.

incoherence to believe that considerations other than the moral status of the fetus are pertinent to the moral quality of the act of abortion. Why isn't this the case, *mutatis mutandis*, for embryo research?

Ellis and Marino acknowledge what the White House working group noted—that the American people have passed an affirmative moral judgment on IVF procedures, the intention of which they generally applaud. But through a double effect, these procedures generate fertilized oocytes that are not immediately implanted in the primary recipient and are unlikely ever to be implanted subsequently. *Ex hypothesi*, the generation of these embryos—which might have been implanted but were not—is delimited by the intention of facilitating otherwise unattainable human procreation. Research on these embryos takes place under the moral aegis of this intention. By contrast, the creation of embryos for research purposes occurs under different moral auspices and summons up the centuries-old fear of the quest for scientific knowledge unchecked by natural or ethical limits.

It is true, as Cardinal Keeler and others have charged, that this conceptual distinction can be erased in practice:

> For if in vitro fertilization clinics may not use federal funds to create "research embryos" solely for the purpose of research, they can easily fertilize more embryos at the outset than are needed for "reproductive" use by infertile couples—thus ensuring that they will have as many embryos as they want for research.[20]

But this is precisely the kind of abuse that could, as the Panel rightly suggested, be addressed through the enhanced oversight that would accompany federal funding of embryo research.

It is revealing that in its final report, the Panel encountered considerable difficulty in justifying the creation of embryos solely for research purposes. After summarizing a range of moral objections to this practice, the Panel keeps its distance from positions that equate the moral status of research-only embryos with IVF embryos and instead offers a purely consequentialist defense: without research-only embryos, various forms of fertility research would be harder (perhaps impossible) to conduct.[21] This line of argument will persuade only those who already accept the idea that in the area of embryo research the end justifies the means. It does not directly engage, or counter, arguments based on moral side-constraints—for example, the claim that the creation of embryos solely for research purposes is "inherently disrespectful of human life."[22]

---

[20] Letter to President Clinton, December 7, 1994.

[21] Report, pp. 54-56.

[22] Report, p. 53.

In the end, we must grapple with a multiplicity of moral considerations, each of which is relevant to embryo research and none of which is dispositive. Claims resting on the moral status of the embryo, on the manner of its coming into being, and on the consequences of conducting (or not conducting) research all have weight. Carried to extremes, moral positions that urge us to ignore consequences (e.g. "Let justice reign, though the earth perishes"[23]) lack credibility. On the other hand, there is a powerful presumption in favor of treating embryos carefully and with respect, as what the Panel itself calls a "developing form of human life." In my judgment, the Panel fails to make a compelling case that the consequences of not creating embryos for research purposes would be so negative as to rebut or override this presumption. But because this judgment rests on moral deliberation—a balancing of incommensurable moral considerations—rather than a logically compelling argument, others may disagree. In this respect (among others), the gap between moral and political decision-making may be narrower than many suppose.

---

[23] A rough translation of the Latin maxim, "Fiat justitia, pereat mundus," quoted frequently by Kant.

HUMAN SACRIFICE AND HUMAN EXPERIMENTATION: REFLECTIONS AT NUREMBERG[1]

Jay Katz

I.

"Life is short, the art long, opportunity fleeting, experiment treacherous, judgment difficult," Hippocrates once said. On the fiftieth anniversary of the Doctors' Trial, which charged Nazi physicians with "crimes against humanity" and violations of Hippocratic ethics in the conduct of human experimentation, I want to begin with Hippocrates' observation that to "experiment [is] treacherous."

Being aware of medicine's limited ability to cure and, thus, the temptation to resort to dangerous, heroic measures, Hippocrates admonished his fellow physicians, "[a]s to disease, make a habit of two things—to help, or at least to do no harm." Hippocrates was not opposed to human experimentation in the practice of medicine, but in his day physicians experimented primarily to benefit individual patients, once customary remedies had proven ineffective.

At the dawn of medical science in the mid-1850s, "experiment treacherous" assumed a dimension not contemplated by Hippocrates. For the first time, experimentation would extend to countless patients, not for their direct benefit, but to advance scientific knowledge for the benefit of humankind. Medicine now held the promise of reversing Hippocrates' aphorism: Life would be longer, art shorter, science longer, opportunity enduring, judgment easier. To accomplish these objectives, a new breed of scientific physician-investigators expected their patient-subjects to make sacrifices on behalf of medical science. Thus, experiment would become even more treacherous.

The philosopher Hans Jonas, in a remarkable essay on human experimentation, comes close to equating human experimentation with the "primeval human sacrifices... that existed in some early societies [for] the solemn execution of a supreme, sacral necessity'; for he suggested that both involved "something sacrificial [in their] abrogation of personal inviolability and the ritualistic exposure to gratuitous risk of health and life, justified by a presumed greater social good." Whatever the relationship between ancient religious practices of sacrifice as an offering to a deity and scientific research practices of sacrifice as an offering to medical progress, the readiness with which human sacrifice for the sake of medical progress has been embraced by the medical

---

1 An earlier version of this paper was published as Yale Law School Occasional Papers, Second Series, Number 2. Used with permission.

profession is remarkable. As one distinguished surgeon put it: "[Conducting] controlled studies may well sacrifice a generation of women but scientifically they have merit."

Rene Girard, in his book *Violence and the Sacred*, observes that "[i]n many rituals the sacrificial act assumes two opposing aspects, appearing at times as a sacred obligation to be neglected at great peril, at other times as a sort of criminal activity entailing perils of equal gravity." The conflict between medicine and law on the permissible limits of human experimentation, to which I shall return repeatedly, reflects these "opposing aspects." When do such "sacred (scientific) obligations" become a "criminal activity"?

Sacrifice can be voluntary or involuntary. The distinction is crucial. But I shall argue that even voluntary sacrifice can be safeguarded only if investigators learn that seeking voluntary consent is their moral obligation, if they learn to desist from employing the concept of voluntary consent as a deceptive subterfuge to shift moral responsibility for participation in research from themselves to their patient-subjects.

In my work I have been largely concerned with involuntary sacrifice, which brings to the surface a conflict inherent in al human research: respect for individual inviolability, on the one hand, and the pursuit of scientific knowledge for the benefit of mankind, on the other. Exploring this conflict in the context of the Nazi concentration camp experiments may seem ludicrous, because the brutality and torture inflicted during these experiments was so immoral that it may seem one need not probe further. Yet, I believe that the doctors' conduct illuminates, with flames from hell, less egregious though still troublesome practices that have stalked human experimentation from its beginnings to this day.

The Nuremberg Code is the one document that seeks in uncompromising language to protect the inviolability of subjects of research. It deserves to be taken more seriously than it has been by the research community. We cannot resurrect the dead, but we can learn from their suffering.

II.

When I received the invitation to speak at Nuremberg, I knew that I had to come. But I did not realize then how painful it would be to reimmerse myself in a history that is so inextricably intertwined with my personal and professional life. For what transpired in Nazi Germany has shaped my life as a person, a physician, and a teacher. In all my work, the disadvantaged in our midst, those stripped of their rights and dignity—the mentally ill, women, children, patients, research subjects—have always been my people.

I was born in Germany—in a small town called Zwickau, Saxony—and lived there until 1938. After a year in Czechoslovakia, my immediate family escaped to England a few weeks before the invasion of Poland. Seven months later we arrived in the United States, and I eventually studied medicine at Harvard Medical School. I was a second-year medical student during the Doctors' Trial, but it was never discussed in any lecture or seminar, even though Harvard was a school that encouraged us to become investigators. Only after I joined the Yale Law School faculty, 39 years ago, did I learn in any depth about the concentration camp experiments. A few years later, thoughts of those experiments led me, joined by many students, to a prolonged exploration of the ethical and legal implications of human experimentation, opening up a field of inquiry then pursued by only a handful of others.

As soon as I decided to go to Nuremberg, childhood memories flooded my mind: listening on the radio to Nazi party rallies where Hitler, Hess, Goebbels and others spoke about my people in contemptuous and threatening ways. I was then a frightened Jewish boy, scared to go to school, where I knew I would be vilified and on a few occasions even beaten. I was angry at my parents for not leaving. They thought it would all blow over; a "Final Solution" was beyond our contemplation.

The nightmare is now past; yet its memories are still alive. During the past few months they have haunted my dreams and during many sleepless nights. And they accompany me as I write.

My problems in this writing remain unresolved. They are embedded in my intent to focus on an aspect of my life's work that began with what I learned about Auschwitz but then went beyond the Nazi horrors, to an exploration of physicians' striking inattentiveness to ethical values in the conduct of human experimentation *before, during and after* the Nazi era. To be sure, at no time in the annals of human experimentation have physicians conducted experiments on humans with the sadism witnessed during the darkness of the Nazi period, where, for example, the death of the subjects was an integral part of the research design.

Thus, in making any comparisons between the Nazi experiments and underlying problems in all research on human subjects, no matter how qualified, in the belief that we must learn from history and that its darkest moments have much to teach us, would I detract from the "uniqueness" of the suffering of the millions who were slaughtered, many with the active collaboration of physicians, and of the thousands who perished in the service of human experimentation? Would I make invidious comparisons between the conduct of Nazi physician-investigators and physician-investigators in the rest of the Western world? I put my questions this starkly because they have haunted me during the past months.

I believe that the concentration camp experiments, which transgressed the last vestiges of human decency, can be located at one end of a continuum, but I also believe that toward the opposite end, we must confront a question still relevant in today's world: How much harm can be inflicted on human subjects of research for the sake of medical progress and national survival? Convictions about the existence of hell can make investigators pause and reflect, as it did during the days of the Cold War, when a few American physician-scientists, while contemplating experiments much less egregious than those conducted by the Nazi physicians, asked: "Are we beginning to behave as they did?"

The concentration camp experiments are embedded in the Holocaust, in what happened to my people, my relatives, Gypsies, homosexuals, political prisoners, and prisoners of war. The confluence of many forces—including biological science and the ideology of the Nazi state—made the Holocaust well-nigh inevitable. And physicians' inattentiveness to the problematic of Hippocratic ethics and its oath, which had served medicine well in the days of the Greeks and throughout the Middle Ages but required a thoroughgoing reappraisal at the dawn of the age of medical science, added its own contributions to the Holocaust and the concentration camp experiments.

Since others have addressed the Holocaust—the murders committed during the selection for death or work—I shall address only the human experimentation aspects of the Holocaust. I do want to underscore, however, that, unlike other historical instances of mass murder, the Final Solution was carried out by doctors whose actions were seen as justifications for the Holocaust and euthanasia as well. How could physicians behave that way? How could healers become murderers?

I have no answers. Nor have I read any that satisfy me. Robert Lifton, in his pioneering book The Nazi Doctors, suggests that an explanation can be found in the psychological principle of "doubling: the division of the self into two functioning wholes, so that a part-self acts as an entire self," an Auschwitz self and a non-Auschwitz self. "The Nazi doctors' immersion in the healing-killing paradox," Lifton says, "was crucial in setting the tone for doubling," leading doctors to "[subvert] medicine from a practice of healing to a science of healing. Nazi medicine was not just corrupted, it was inverted."

"Doubling," however, is an all too human phenomenon. Indeed, it is a ubiquitous manifestation of human beings' conflictual nature. And physician-investigators are particularly susceptible to the perils of doubling. In their scientific pursuits doctors are double agents, because their commitment to the objective imperatives of the research protocol conflicts with, and can take precedence over, the individual needs of patients. Thus, in human research, the healing-harming-(killing) "paradox" is inherent in the task itself.

Let me note only in passing that with regard to euthanasia, "the healing-killing paradox" is graphically illustrated in an article published in 1941 in the *American Journal of Psychiatry* by a Cornell Medical School professor, who recommended that "hopelessly unfit children—nature's mistakes—should be killed, and the less unfit [sterilized]" so that "thereafter civilization will pass on and on in beauty." The Nazis began by killing their own "defectives" and then went on to killing Jews and Gypsies, whom they also considered biologically "defective."

I continue to find it inexplicable, despite the many explanations that have been advanced, that involuntary sacrifice, with physicians' active participation or passive acquiescence, went so totally out of control at Auschwitz. Can one say more than Erasmus did? *Homo homini aut deus aut lupus*—Man is to man either a god or a wolf. Do we romanticize physicians too much wen we wish to exclude them from Erasmus' dictum? Must we recognize, for the sake of the future, that the ingredients for what happened at Auschwitz are inherent in the conduct of research and that we must learn to control it better at its source?

Let me return to the Doctors' Trial. I shall relate it in two parts. First, I will describe two experiments most briefly; and then, after a few comments on the history of medical ethics, I will analyze the tribunal's judgment and its implications for the future conduct of human research.

The Doctors' Trial was the first of twelve trials that followed the Nuremberg trial of the major war criminals by the International Allied Military Tribunal. Conducted by American judges, the Doctors' Trial focused on experimentation on human beings during the Nazi regime. Evidence on the experiments was presented over many months in excruciating detail. I have reviewed the record many times and still find it devastatingly painful to read.

Most notorious among the experiments was Dr. Sigmund Rascher's work on the effects of high altitude on human survival. On May 15, 1941, Rascher wrote to Heinrich Himmler:

> [During] a medical selection course [in which] research on high altitude flying played a prominent part [we learned that English fighter planes were able to reach higher ceilings than we could]. Regret was expressed that no experiments on

human beings have so far been possible... because such experiments on human beings are very dangerous and nobody is volunteering. I therefore put the serious question: Is there any possibility that two or three professional criminals can be made available for these experiments?... The experiments in which the experimental subject of course may die... are absolutely essential... and cannot be carried out on monkeys, because monkeys offer entirely different test conditions....

Dr. Rudolf Brandt, on behalf of Himmler, responded promptly: "I can inform you that prisoners will, of course be gladly made available for the high-flight researches. ... I want to use the opportunity to extend to you my cordial wishes in the birth of your son...."

In Raschler's report on one of these experiments, he described in graphic detail

the fate of a 37-year-old Jew in good general condition who, at ever increasing altitudes, began to perspire, to wiggle his head, [and to suffer from severe] cramps. Breathing increased in speed and [he] became unconscious.... Severest cyanosis developed... and foam appeared at the mouth. After breathing had stopped [an electrocardiogram] was continuously written until the action of the heart had come to a complete standstill. One half hour [later] dissection was started.

The freezing experiments, many fatal, were even more brutal, if that is possible. The subjects were immersed in ice water for hours on end. They pleaded to be shot to escape their unbearable agony. As I read these accounts, I could almost hear their pleas. These and the many other experiments conducted at Auschwitz and elsewhere bear testimony to the brutality inflicted on "lives not worth living" and therefore expendable.

Rascher, in his report, was delighted that the heart actions he had recorded "will [prove to be of] particular scientific interest, since they were written down with an electrocardiogram to the very end." For him the experiment represented another triumph in the 100-year history of human sacrifice for the sake of the advancement of knowledge.

Experimentation with human beings antedates the Nazis, of course. Its roots go back to antiquity, but in the 1850s, human research increased in magnitude unprecedented during the millennia of medical history. Academic physicians observed with envy the discoveries in physics and chemistry that had resulted from systematic objective investigations, and they adopted the methodologies of the physical sciences so that medicine would also become a respected scientific discipline. At the same time, doctors lost sight of the fact that it is one thing to experiment with atoms and molecules, and quite another to do so with human beings. Once, while reflecting on the inhumanity of Auschwitz, my thoughts took me back to these beginnings of medical research. I was struck by how quickly physicians accepted these new ways of conducting research with human beings, never asking whether fellow human beings, particularly patients, should be subjected to these novel practices and, if so, with what safeguards.

The initial advances in knowledge that resulted from such scientific investigations, which promised to alleviate human suffering to an extent previously unknown, seemed to justify the means employed. The uncharted moral path led only once to Auschwitz; yet, on many other occasions down the road, human beings would pay a considerable price for the sake of medical progress.

The early fruits of medical research were spectacular. The bacterial etiology of many diseases was proved, resulting in cures never before the lot of humankind. Investigations of the use of x-rays to see the previously invisible revolutionized diagnostic techniques. Experiments with various anesthetic agents led to remarkable advances in surgery.

These experiments were largely conducted in public hospitals with the poor, with children, women, prostitutes, the elderly—that is, with the disadvantaged and the downtrodden. Albert Moll, in his remarkable book *Arztliche Medizin*, published in 1902, described many experiments conducted with patient-subjects throughout Europe and the United States during the late nineteenth century. He was particularly troubled by experimentation with the terminally ill, who frequently served, as they still do, as subjects of research. Since they would soon die anyway, learning from them seemed self-evidently the right thing to do. In reading these accounts my mind turned again to the Auschwitz subjects, who were also terminal cases—"lives not worth living"—soon to be reduced to ashes.

Human research and its contributions to the advancement of knowledge captured the imagination of doctors. The promise that omnipotence would replace the earlier struggle against impotence, and the promise of fame, academic advancement, and perhaps even economic fortune, loomed large in physicians' minds.

But the intrusion of research into the clinical practice of medicine required keeping the two enterprises separate. Patients went to doctors to be helped and not to serve as research subjects. Crucial distinctions needed to be made between clinical care and human research for the advancement of science. Instead, the boundaries between therapy and research became blurred. The "therapeutic illusion"—that research would in some undefined ways benefit subjects—contributed to this obfuscation.

Although physician-investigators were aware of the pain suffered by, and the occasional deaths of, their patient-subjects, they did not consider whether they might be violating their Hippocratic duty not to harm their patients. I shall return to this problem shortly. For now, let me note that this history reveals antecedents to the concentration camp experiments. However different they were in degree of torture and brutality, the experiments conducted by pre-World War II physician-investigators, largely medical school professors, were precursors to what transpired at Auschwitz. Medical students observed their teachers, read about their scientific investigations and their uses and abuses of patients. Dr. Helmut Poppendick, one of the Nazi doctor defendants, put it this way: "I knew [from my student days] that the modern achievements of medical science had not been brought about without sacrifices." Would the Nazi doctors have behaved differently without that history? At least some of them might have paused and reflected had they been differently educated.

When medical science and medical practice became intertwined, a new ethical question should have been raised: Are physicians' obligations to their patient-subjects different from their obligation to their (other) patients? But only a few remarkable physicians considered that question, and their concerns were not heeded.

III.

So far I have focused on the beginnings of objective medical science. I now want to return to the history of medical authoritarianism. The combination of the relentless pursuit of science, with its inherent dangers of objectifying subjects, became embedded in the ancient tradition of medical authoritarianism, with its objectification of patients. Both dynamics make it difficult to respect patient-subjects as persons with their own interests and rights.

In my readings on medicine from ancient and medieval times up to the present, I was impressed by physicians' awareness of their relative impotence, on the one hand, and their conviction that they had something useful to offer, on the other. In the late seventeenth century, the physician Samuel de Sorbiere wrote that medicine "is a very imperfect science, quite full of guesswork, and ... scarcely [understands] its subject matter." He was of two minds about whether to be truthful with patients or to foster their unconditional confidence in their physicians because such confidence served the purposes of cure.

This conflict was generally resolved by encouraging physicians to be authoritarian—to demand that patients be obedient and follow doctors' orders if they wanted to be helped. "Should the patient not submit to your discipline… do not persevere in the treatment," said the physician Isaac Israeli, around 900 A.D. And the surgeon Henri de Mondeville wrote around 1200 A.D., "the surgeon should promise that if the patient will obey the surgeon… he will soon be cured." As late as the mid-1950s, the influential sociologist Talcott Parsons observed: "[T]he doctor-patient relationship has to be one involving an element of authority—we often speak of 'doctors' orders'."

Any disclosures were limited to enlisting patients' cooperation; otherwise, as Hippocrates had put it, "[reveal] nothing of the patient's future or present condition." A patient's blind trust was considered essential, even though, beyond comforting attention and a few potions, physicians had little to offer for the cure of disease.

In sum: Two precepts were handed down from generation to generation of Hippocratic physicians: to avoid doing harm and to insist on silent obedience. The latter, in particular, had far-reaching consequences for the conduct of research; for the same authoritarianism with which patients had traditionally been treated in therapeutic settings was imposed on subjects of research, who learned little, if anything, about the scientific purposes for which they were recruited.

Physician-investigators seemed oblivious to these moral issues and, if they were not, charged ahead anyway. Ultimately—and, ironically, first in Germany—the state took notice and began to regulate research. In 1900, Dr. Albert Neisser was put on trial after it became known that he had injected serum from patients with syphilis into patients, largely prostitutes, suffering from other diseases. The German academic medical community defended his conduct. Lawyers, on the other hand, argued that "non-therapeutic research without full consent fulfilled the criteria of physical injury in criminal law."

That same year, the Prussian Parliament enacted the first, limited state regulation of research. The lawyer Ludwig von Bar, a consultant to the Prussian Minister, put it well: "Respect for rights and morality has the same importance for the good of mankind as medical and scientific progress." One hundred years later, his assertion is still being contested, and it was most flagrantly disregarded by the Nazi doctors.

In 1931, the Weimar Republic enacted regulations providing protection not only for subjects of non-therapeutic research but also for patients receiving "innovative therapy." The minutes of the 1930 meeting that preceded the enactment of these regulations record how the academic physicians made light of what they called "rare" abuses, emphasizing instead the importance of advancing medical science. Only

Dr. Julius Moses, a physician and member of the German Reichstag, argued for official guidelines to protect patients from dangerous experiments. His was a lone voice among physicians.

The role of the state in the regulation of medicine raises complex questions that I cannot discuss in this piece. Note, however, that in the United States common law judges, not physicians, promulgated the doctrine of informed consent in medical practice, giving some decision-making authority to patients. Medicine has never created a regulatory framework for its practices, and this lack of a structure may come to haunt us in this age of managed care and physician-assisted suicide.

Michael Kater concluded his scholarly analysis of Doctors Under Hitler with the observation that

[I]t was in the interpersonal relationship between healer and patient that

German medicine corrupted itself [by contravening] the most important principle of the Hippocratic Oath... 'I will use treatment to help the sick according to my ability and judgment, but never with a view to injury and wrongdoing.'

But as I have tried to demonstrate, the corruption Kater speaks of has a long history. Moreover, the Hippocratic Oath—a document that emphasizes physicians' obligation of caring attention toward individual patients—says nothing about the ethics of human research that has relevance for an age of scientific medicine, unless one wants to invoke the oath to put a stop to most research. Thus, after the dawn of the age of medicine, the medical profession failed its members and its patients by not modifying its oath to reconcile its commitment to patients' welfare with radically changed circumstances.

For many reasons, physicians have preferred to view human experimentation merely as an extension of medical practice. In 1916 the Harvard physician Walter Cannon recommended to the House of Delegates of the American Medical Association that it endorse the importance of obtaining patient consent and cooperation in human experimentation. His proposal, however, was not brought up for consideration. One influential physician remarked, "It would open the way for a discussion of the importance of obtaining the consent of the patient before any investigations are carried on which are not primarily for the welfare of the patient."

And this is only half of the story. Disclosure in these contexts would require discussions with patient-subjects of the uncertainties inherent in therapeutic medicine as well; and, if that were to happen, the question would arise: Why should not patients be similarly informed? Physicians feared that their authority to make decisions on behalf of patients would be undermined and patients' best interests would be detrimentally affected. Doctors viewed such prospects, as they still do, as a threat to the traditional practice of medicine. They valued silence, their own and their patients', for silence maintains authority.

IV.

The Doctors' Trial confronted the tribunal with aberrational, almost unbelievable, accounts of what physician-scientists can do and justify when respect for human dignity is totally abrogated. In its final judgment the tribunal went beyond these facts and articulated a vision of the limits of scientific medical research that was clear and unambiguous. To be sure, its pronouncement would eventually require elaboration and modification, but it was the uncompromising clarity of its vision about the primacy of consent that proved so disturbing to the medical community.

The message of the tribunal might easily have been blunted by the confusing or inaccurate allegations made throughout the trial. In his opening statement Telford Taylor, then chief counsel for the prosecution of war crimes, charged the doctor-defendants "with murders, tortures and other atrocities committed in the name of medical science." But in his closing argument, James McHaney, the chief prosecutor for the medical case, redirected the tribunal's attention to what he considered the nub of the case:

> [T]hese defendants are, for the most part, on trial for the crime of murder.... It is only the fact that these crimes were committed in part as a result of medical experiments on human beings that makes this case somewhat unique. And while considerable evidence of a technical nature has been submitted, one should not lose sight of the true simplicity of this case.

Thus, was it a murder trial of ordinary criminals, who also happened to be doctors, or of medical scientists (and medical science) whose conduct made them murderers? The ensuing and prolonged disregard of the Nuremberg Code by members of the medical profession depended on their answer to this question. Most focused on the barbarism of the Nazi doctors' conduct and concluded that the code was relevant only to Nazi practices—not to research in a civilized world. They disregard the fact that murder and torture were not to sole issues before the court, that the permissible limits of scientific research were on trial as well.

The tribunal addressed both issues: "War crimes and crimes against humanity" and rules that must be observed in the conduct of medical experimentation. With respect to the latter, the tribunal observed that "medical experiments...when kept within reasonably well defined bounds conform to the ethics of the medical profession." But then judges immediately asserted that "[a]ll agree, however, that certain basic principles must be observed in order to satisfy moral, ethical and legal concepts."

The phrase "all agree" was confusing. Who were these "all"? Surely not the Nazi doctors, among them some of the most distinguished German medical scientists, and surely not many physician-investigators of the nineteenth and twentieth century. Nor do many contemporary medical scientists embrace the tribunal's principles. The confusion, I believe, had its origins in the previously noted disagreement over the issues that required adjudication: All agreed with the prohibition against murder and torture; but "all" did not agree with the tribunal's "basic principles" for the conduct of research.

V.

Of the ten principles known as the Nuremberg Code, the first will be my focus here. It reads, "The voluntary consent of the human subject is absolutely essential." The judges did not, however, stop there. Instead they went to unusual lengths to define voluntary consent, in terms of both subjects' capacity to give consent and the information that investigators must provide to subjects. It is the detailed disclosure requirements which, I believe, the research community has found difficult to accept.

The judges wondered whether they had gone too far in imposing their legal views on the medical profession: "Our judicial concern, of course, is with those requirements which are purely legal in nature.... To go beyond that point would lead us into a field that would be beyond our sphere of competence." But if indeed they did venture "beyond [their] sphere of confidence," they were compelled to do so. Whatever their ignorance of medicine's needs, being American judges—steeped in the self-determination ideal, so much celebrated in our political tradition—they wanted their first principle to safeguard human dignity and inviolability, in research and civilized life.

The judges then shifted their focus back to the concentration camp experiments. This, too, proved confusing, because it created the impression that their code applied only to the case before them, that they were not addressing the entire universe of human experimentation. Yet, in their preamble to the Nuremberg Code, they had suggested otherwise. There they spoke to this entire universe when they averred that they wanted to promulgate "basic principles [that] must be observed in order to satisfy moral, ethical and legal concepts [in] the practice of human experimentation." This is my view of their intent. And this gives their pronouncements historic significance for the post-Nuremberg conduct of experimentation with human beings.

In the most uncompromising language, the judges suggested in their first principle that the tensions between progress in medical science and the inviolability of research subjects must be resolved in favor of respect for the person, his or her self-determination and autonomy. Consent became the necessary justification for the conduct of research, though not a sufficient justification. In most of its nine other principles, the tribunal spelled out other conditions that must be met before human beings could even be asked to serve as means for others' ends. These conditions include importance of the research question, prior animal experimentation, and avoidance of unnecessary or predictably disabling injury or death.

Critics have correctly observed that the first principle was irrelevant to the case before the tribunal, for the basic problem with the concentration camp experiments was not that the subjects did not agree to participate; it was the brutal and lethal ways in which they were used. But, in my view, the first principle, like the rest of the code, did not speak to what transpired at Auschwitz; it spoke to the future. The fact that with regard to Auschwitz the first principle was indeed irrelevant (almost silly), is evidence of the judges' broader objective. American judges are not averse to going beyond the facts of a case; in this instance, I am glad that they did.

Another confusion was introduced by the medical experts for the prosecution, who asserted that in the rest of the Western world, physician-investigators conducted their research according to the highest ethical medical standards, including obtaining consent. I doubt that the judges believed them. On cross-examination, Dr. Andrew C. Ivy was forced to admit that the first written AMA code on human experimentation was enacted while the trial was under way, a fact he had tried to hide on direct examination. Whatever the judges' reaction to this testimony, they required little convincing that physician-investigators should not use human beings for research without consent and, if they had done so in the past outside of Auschwitz, such practices should cease. Their convictions on that point were only reinforced by the nightmarish stories they had just heard.

They could not know that for decades their code would make little impact on research practices; that many violations would continue to occur in the United States and elsewhere. For example, the Tuskegee Syphilis Study, in which the lives and health of many African-Americans were ruined, was not stopped until 1972. That study had been conducted by the U.S. Public Health Service with 400 uninformed African-American men in order to gather data on the natural history of untreated syphilis from its inception to death. The study should never have begun, and it surely should have been stopped in the early 1940s when elective treatment for some of the late manifestations of syphilis became available.

Or consider the experimental injection of live cancer cells into uninformed elderly patients at the Brooklyn Jewish Chronic Disease Hospital. Or consider the experimental injection of plutonium into uninformed pregnant women to learn whether plutonium crosses the placental barrier, conducted at a time when little was known about plutonium and its dangers. Or consider the total-body radiation experiments with terminally ill patients at the Cincinnati University Hospital. The plutonium and radiation experiments were conducted during the Cold War and were "justified" on grounds of national defense, an argument that had also been used by the Nazi physicians for what they had done. Finally, consider more recent drug studies to determine the toxicity of new cancer treatments, which were presented to patient-subjects not as research but as "new and promising frontier treatments."

These experiments were not comparable to the Nazi research, for care was always taken to keep physical harm to a minimum, but neither so they meet the standards of the Nuremberg Code. As one American research scientist put it, "I am aware of no investigator (myself included) who was actively involved in research…in the years before 1965 who recalls any attempts to secure 'voluntary' and informed consent according to Nuremberg's standards.

In giving preeminence to "voluntary consent" in the conduct of research, the judges sought to admonish investigators to become more respectful of subjects' dignitary interests in making their own decisions in interactions with investigators. Implementation of that objective remains the unfinished legacy of the Nuremberg judges. For the regulations that now require consent will not adequately protect the rights of subjects to self-determination unless physician-investigators embrace these rights as a new Hippocratic commitment.

Vulnerable subjects are compelled by their necessitous circumstances to place their trust in physicians whom they consider care givers, not investigators. The problem of "trust" surfaced in one of the studies conducted by the President's Advisory Committee on Human Radiation Experiments during the Cold War, in which we assessed attitudes toward research among many hundred of patient-subjects who as recently as 1994 were enrolled in research projects. We discovered that patient-subjects believed that "an [experimental] intervention would not even be offered if it did not carry some promise of benefit for them," and that therefore the consent process was "a formality" to which they need not give much thought.

The lesson to be learned from our findings is clear: Consent will never be truly informed or voluntary unless patient-subjects are disabused of that belief. Their rights can be protected only if physician-investigators acknowledge that their patient-subjects view them as physicians and not investigators, and that they, the doctors themselves, have the responsibility not to betray that trust in research settings. Patient-subjects must be told that their own and their physician-investigators' agendas are not the same. Research is not therapy.

This is a formidable undertaking and a consequential one, about which I have written extensively. It takes time, may impede research because of too many refusals, and may thereby make some experiments impossible to conduct. Choices have to be made between the relentless pursuit of medical progress and the protection of individual inviolability. The latter, however, will be given the weight it deserves only if doctors learn to respect patient-subjects as persons with minds of their own and with the capacity to decide for themselves how to live their medical lives. Their choices may or may not include a willingness for altruistic self-sacrifice, but such choices must take precedence over the advancement of science.

These are the conclusions to which my work has brought me. It all began with reading about Auschwitz, which led me on a long journey, during which I learned much about what human beings can do to one another in less egregious though still painful ways. Without my and my people's past, I might never have embarked on that journey.

VI.

In conclusion, I return to questions I raised somewhat differently at the beginning of this essay: Am I doing justice or injustice to the victims of the concentration camp experiments by placing their suffering in the context of the historical processes by which they came about? Am I doing an injustice to the victims by comparing their fate with that of other research subjects whose lives were not reduced to ashes? Am I doing them an injustice, since violent death is always a tragedy, to celebrate the Nuremberg Code that resulted from their suffering?

What is justice, what is injustice? A friend of mine once pointed out to me the repetition of the word justice in Deuteronomy: *"Tzedek, tzedek tirdof"* (Justice, justice, shalt thou pursue). Such a seemingly unnecessary repetition always invites commentary, and the one he heard was this: "Justice can never be adequately pursued only as a goal or an idea; it is also reflected in the means employed."

I have attempted to employ the proper means, by comparing this tragic episode with its past as well as with the present. And I have attempted to pursue justice—to do justice to the victims—not merely by commemorating their suffering but also by construing the Nuremberg Code as their unwitting legacy. They were subject to coercion, sadism and torture; the Nuremberg Code celebrates freedom and human dignity.

As medical professionals, we remain unconvinced that we should embrace the Code's principles in the spirit in which they were promulgated. It remains my dream that we shall do so. It may only be a dream, but it comforts my nightmares.